T0210768

New Aspects of Quantity Surveying Practice

The construction industry is undergoing great change particularly with the introduction of digital technologies and the increasing emphasis on sustainability and ethical practice. The fifth edition of *New Aspects of Quantity Surveying Practice* introduces and discusses these changes and their impact on the industry.

The book champions the adaptability and flexibility of the quantity surveyor, whilst covering the hot topics which have emerged since the previous edition's publication, including:

- A new chapter on the impact of digital construction
- Sustainable construction
- Procurement trends
- Ethics and ethical practice
- The RICS Futures (2020) publication

The book is essential reading for all quantity surveying students, teachers and professionals. It is particularly suited to undergraduate professional skills courses and non-cognate postgraduate students looking for an up to date understanding of the industry and the role.

Duncan Cartlidge FRICS is a Chartered Surveyor with extensive experience in the delivery and management of built assets, as well as providing education and training to a wide range of built environment professionals and contractors. He is the author of several best-selling books, including *The Quantity Surveyor's Pocket Book* (4th edition), *The Construction Project Manager's Pocket Book* (2nd edition) and *The Estimator's Pocket Book* (2nd edition). www.duncancartlidgeonline.com

New Aspects of Quantity Surveying Practice

Fifth Edition

Duncan Cartlidge

Routledge
Taylor & Francis Group

LONDON AND NEW YORK

Designed cover image: © Shutterstock

Fifth edition published 2023
by Routledge
4 Park Square, Milton Park, Abingdon, Oxon OX14 4RN

and by Routledge
605 Third Avenue, New York, NY 10158

Routledge is an imprint of the Taylor & Francis Group, an informa business

© 2023 Duncan Cartlidge

[First edition published by Routledge]
[Fourth edition published by Routledge]

British Library Cataloguing-in-Publication Data
A catalogue record for this book is available from the British Library

Library of Congress Cataloguing-in-Publication Data
Names: Cartlidge, Duncan P., author.
Title: New aspects of quantity surveying practice / Duncan Cartlidge.
Description: Fifth edition. | Abingdon, Oxon ; New York, NY :
Routledge, 2023. | Includes bibliographical references and index. |
Summary: "The construction industry is undergoing great change
particularly with the introduction of digital technologies and the increasing
emphasis on sustainability and ethical practice. The fifth edition of New
Aspects of Quantity Surveying Practice introduces and discusses these
changes and their impact on the industry. The book is essential reading for
all Quantity Surveying students, teachers and professionals. It is particularly
suited to undergraduate professional skills courses and non-cognate
postgraduate students looking for an up to date understanding of the
industry and the role"-- Provided by publisher.
Identifiers: LCCN 2022047123 | ISBN 9781032275963 (hbk) |
ISBN 9781032275956 (pbk) | ISBN 9781003293453 (ebk)
Subjects: LCSH: Quantity surveying.
Classification: LCC TH435 .C365 2023 | DDC 692/.5--dc23/eng/20221012
LC record available at https://lccn.loc.gov/2022047123

ISBN: 978-1-032-27596-3 (hbk)
ISBN: 978-1-032-27595-6 (pbk)
ISBN: 978-1-003-29345-3 (ebk)

DOI: 10.1201/9781003293453

Typeset in Bembo
by MPS Limited, Dehradun

Contents

Figures

Tables

Foreword

The construction industry has long understood that it needs to transform its performance to ensure it is ready and able to meet the changing needs of its clients, now and in the future. This means becoming more diverse to increase the capability and capacity across its workforce; harnessing new technology to drive efficiency and productivity; becoming a leader rather than a laggard in decarbonisation; and delivering much greater social, environmental and economic impact in every project and programme.

Published nearly a decade ago, the UK government's Construction 2025 vision for the industry clearly conveyed these principles. Perceptions may have been gradually changing on construction services and outputs, but since the turn of the millennium, the reality is that the country's construction industry has not just lagged other sectors such as manufacturing, production and services, the Office for National Statistics, Productivity in the construction industry, UK: 2021 shows that productivity growth has actually been slower. Despite this, construction showed itself as adept as any industry in response to the COVID-19 pandemic, continuing to service projects by rapidly adapting to new, digital ways of working.

The case for reinvigorating construction, not just in the UK but globally, has now shifted from a clear to an urgent need with the public and private sector recognising the critical role it has to play in tackling societies biggest challenges – from climate change to deep-rooted inequalities.

The opportunity for industry could not be bigger, and the same applies for our present and future professionals. It is a particularly exciting time for quantity surveyors because the use of digital tools means we have more capacity, and will be better equipped, to provide clients with insight and analysis that transforms the performance of their projects and programmes.

Relevant cost data has always been at the heart of robust cost advice. The future quantity surveyor will have even bigger and better cost data at their fingertips. The management of big data will be critical, particularly with local silo service delivery reducing and the capacity of the construction industry to deliver globally through digital innovation.

Projects and programmes are not just delivered in small, isolated markets any longer because materials, and labour, come from all over the world. It is imperative that we are able to advise clients on what impact economic headwinds, supply chain disruption or geo-political tensions will have on cost and delivery. The right data gives us the ability to do that, and we must embrace technology to give us the time to forecast how the markets changes are affecting price, how clients need to purchase.

The shift to carbon management is also seismic for our profession. We now have the tools to provide consistent and accurate assessments of a project's embodied carbon count from an early design stage. While there will be further advances in the future, the ability to do this has a vital part to play in clients' journeys towards a robust, measurable net zero ambition. Unlocking innovation like this is key to ensuring that our industry is part of the solution to tackling the pressing social, environmental and economic challenges we face.

As we look to the future, we have to ensure that carbon management and digital service delivery are at the heart of what we do. Quantity surveyors need to approach their role with this mindset, and the latest edition of *New Aspects of Quantity Surveying Practice* lays this out with exacting clarity. By building more diverse talent pools and embedding this approach and these skills into professional pathways from day one, we can transform our industry to create a green, inclusive and productive world.

Martin Sudweeks
UK Managing Director, Turner and Townsend Cost Management

Preface to first edition

The Royal Institution of Chartered Surveyor's Quantity Surveying Think Tank: *Questioning the Future of the Profession*, heard evidence that many within the construction industry thought Chartered Quantity Surveyors were: arrogant, friendless and uncooperative. In addition, they were perceived to; add nothing to the construction process, failed to offer services which clients expected as standard and too few had the courage to challenge established thinking. In the same year Sir John Egan called the whole future of quantity surveying into question in the Construction Industry Task Force report Rethinking Construction and if this weren't enough; a report by the University of Coventry entitled 'Construction Supply Chain Skills Project', concluded that quantity surveyors are 'arrogant and lacking in interpersonal skills'. Little wonder then that the question was asked 'Will we soon be drying a tear over a grave marked "RIP Quantity Surveying, 1792–2000"?' Certainly the changes that have taken place in the construction industry during the past 20 years would have tested the endurance of the most hardy of beasts. Fortunately, the quantity surveyor is a tough and adaptable creature and to quote and paraphrase Mark Twain 'reports of the quantity surveyors death are an exaggeration'.

I have spent the past thirty years or so as a quantity surveyor in private practice, both in the UK and Europe, as well as periods as a lecturer in higher education. During this time I have witnessed a profession in a relentless search for an identity, from quantity surveyor to: building economist, to construction economist, to construction cost advisor, to construction consultant, etc. etc. I have also witnessed and been proud to be a member of a profession that has always risen to a challenge and has been capable of reinventing itself and leading from the front, whenever the need arose. The first part of the twenty first century holds many challenges for the UK construction industry as well as the quantity surveyor, but of all the professions concerned with the procurement of built assets, quantity surveying is the one that has the ability and skill to respond to these challenges.

This book, therefore is dedicated to the process of transforming the popular perception that, in the cause of self preservation, the quantity surveyor is wedded to a policy of advocating aggressive price led tendering with all the problems that this brings, to one of a professional who can help deliver high value capital projects on time and to budget with guaranteed life cycle costs. In addition it is hoped that this book will demonstrate beyond any doubt that the quantity surveyor is alive and well, adapting to the demands of construction clients and what's more, looking forward to a long and productive future. Never the less, there is still a long hill to climb. During the production of this book I have heard major construction clients call the construction industry – 'very unprofessional' and the role of the quantity surveyor compared to that of a 'post box.'

In address to the Royal Institution of Chartered Surveyor in November 2001, the same Sir John Egan, that had called the future of the quantity surveyor into question, but now as Chairman of the Egan Strategic Forum for Construction, suggested that the future for Chartered Surveyors in construction was to become process integrators, involving themselves in the process management of construction projects and that those who clung to traditional working practices faced an uncertain future. The author would whole heartedly agree with these sentiments.

'The quantity surveyor is dead – long live the quantity surveyor – masters of the process!'

Duncan Cartlidge
www.duncancartlidge.co.uk

Preface to the second edition

Four years have passed since the first edition of *New Aspects of Quantity Surveying Practice*. At that time 'Building', the well known construction industry weekly, described quantity surveying as *'a profession on the brink'* whilst simultaneously forecasting the imminent demise of the quantity surveyor and references to; *'Ethel the Aardvark goes Quantity Surveying'*, had everyone rolling in the aisles. In a brave new world where confrontation was a thing of the past and where the RICS tried to deny quantity surveyor's existed at all, clearly they was no need of the profession! But wait; what a difference a few years can make for on 29 October 2004 the same publication that forecast the end of the quantity surveyor had to eat humble pie when the Building editorial announced that *'what quantity surveyors have to offer is the height of fashion – Ethel is history'*. It would seem as if this came as a surprise to everyone, except quantity surveyors!

Ironically, in 2006 quantity surveyors are facing a very different challenge to the ones that were predicted in the late 1990's. Far from being faced with extinction the problem now is a shortage of quantity surveyors that has reached crisis point, particularly in major cities like London. The *'mother of all recessions between 1990–1995'* referred to in chapter one had the effect of driving many professionals, including quantity surveyors out of the industry for good, as well as discouraging school leavers thinking of embarking on surveying degree courses. As a consequence there now is a generation gap in the profession and with the 2012 London Olympics on the horizon, as well as buoyant demand in most property sectors, many organisations are offering incentives and high salaries to attract and retain quantity surveying staff. In today's market place a 'thirty something' quantity surveyor with ten to fifteen years experience is indeed a rare, but not endangered, species. It would also seem as though the RICS has had second thought about the future of the quantity surveyor. A survey carried out by the Royal Bank of Scotland in 2005 indicated that quantity surveyors are the best paid graduate professionals. In November 2005, the RICS announced that, after years of protest, the title quantity surveyor was to reappear as an RICS faculty as well as the RICS website.

The new millennium found the construction industry and quantity surveying on the verge of a brave new world – an electronic revolution was coming, with wild predictions on the impact that IT systems and electronic commerce would have on the construction industry and quantity surveying practice; the reality, is discussed in Chapter 5.

For the quantity surveyor, the challenges keep on coming. For many years the UK construction industry has flirted with issues such as whole life costs and sustainability/green issues; it now appears that these topics are being taken more seriously and are discussed in Chapter 3. The RICS Commission on Sustainable Development and Construction recently developed the following mission statement; '*To ensure that sustainability becomes and remains a priority issue throughout the profession and RICS*' and committed itself to raising the profile of sustainability through education at all levels from undergraduate courses to the APC. In the public sector, the new Consolidated EU Public Procurement Directive due for implementation in 2006 now makes sustainability a criteria for contract awards and a whole raft of legislation due in the Spring 2006 has put green issues at the top of the agenda. Links have now been proved between the market value of a building and its green features and related performance.

Following the accounting scandals of the Enron Corporation in 2003 quantity surveyors are being called on to bring back accountability both to the public and private sectors and world wide expansion of the profession continues with further consolidation and the emergence of large firms moving towards supplying broad business solutions tailored to particular clients and sectors of the market.

Where to next?

Duncan Cartlidge
www.duncancartlidge.co.uk

Preface to third edition

Quantity surveying remains a diverse profession with surveyors moving into new areas, some of which are outlined in chapter seven.

The Preface to the second edition of New Aspects of Quantity Practice referred to the increasing interest from both the profession and the construction industry in sustainability and green issues. During the past five years, since the previous edition, sustainability has risen to world prominence and the construction industry, worldwide, has been identified as No 1 in the league table of pollutes and users of diminishing natural resources. It is unsurprising therefore that sustainability has risen to prominence in the industry with many undergraduate and post-graduate programmes now including dedicated modules on sustainable development and clients, professionals, developers and contractors seeking to establish their green credentials.

Ethics, both personal and business and professional standards have also risen to prominence. Never before has the behaviour of politicians, public figures and professionals been under such close scrutiny, the age of transparency and accountability can truly said to have arrived. Although ethics has a long history of research and literature in areas such as medicine, the amount of guidance available for surveyors has been almost non-existent until recently and even now cannot be described comprehensive.

One thing that has been a common theme throughout the writing of the three editions of this book is that quantity surveyors feel unloved not least by their professional institution, the RICS. In 2010 quantity surveyors, not for the first time, threatened to leave the RICS in response to the introduction of AssocRICS, a new grade of membership that it was thought, would result in a lowering of entry standards to the institution.

Layered on top of the above is what has been described as, the deepest recession since the 1920s, with all the challenges that this brought. As this book goes to press, it is still uncertain how many large public sector projects will be axed as the aftermath of the credit crunch lingers on and continues to impact on the construction industry's order books.

Never-the-less, despite world recessions and new areas of focus for practice, the quantity surveyor continues to prosper, with interest in the profession never higher. Therefore, raise a glass to the quantity surveyor, by any definition, a true survivor.

Duncan Cartlidge
www.duncancartlidge.co.uk

Preface to fourth edition

A common theme running through the prefaces to the three previous editions of *New Aspects of Quantity Surveying Practice* was that quantity surveyors were under threat of extinction and it will be no great surprise to learn that little has changed with the fourth edition; this time the menace lurking in the shadows to wipe quantity surveyors of the face of the earth is Building Information Modelling (BIM). The latest indications are the introduction of BIM into UK construction is stalling, for a number of reasons outlined in Chapter 4, never-the-less quantity surveyors cannot afford to ignore the trend to replace drawn information with digital outputs and adapt accordingly.

Quite rightly ethics and ethical practice continues to gain a much higher profile in the everyday life of the construction industry and allied professions. For too long construction's image has been blackened with rumours of bribery, bid rigging and dodgy deals and now clear guide lines are available and are discussed in Chapter 5.

Governments on both sides of the border still appear to be convinced that they are receiving poor value for money from the UK construction industry and continue to publish what seems to be an endless stream of reports and targets some of which are addressed in this fourth edition.

On June 23rd 2016, it was announced that the United Kingdom had voted to leave the European Union. At the time of writing the fourth edition of this book there is little clarity on what BREXIT will actually mean for the UK's relationship with the EU and it is unclear what level of access the UK and in particular UK construction, will continue to have to the EU market. For example, if the UK were to adopt the so-called Norwegian model it would become a member of EFTA and the EEA and the UK's obligations as members of the EEA, would include the adoption of the EU procurement rules. Therefore, for this fourth edition the references and sections on EU public procurement has been retained.

Finally, it is pleasing to report that after years of discontent, peace appears to have broken out between quantity surveyors and the RICS, long may it last.

Duncan Cartlidge
www.duncancartlidge.co.uk

Preface to fifth edition

Brave in our thinking and collaborative in our approach.

The period since the publication of the fourth edition of *New Aspects of Quantity Surveying Practice* and this new edition have been very eventful. The following aspects are reflected in the fifth edition.

BREXIT became a reality leaving in its wake:

* rising costs of construction
* skilled labour shortages and
* excessive lead times.

Just as the ink had dried on the BREXIT agreement, the world was hit by a pandemic that caused major disruption with:

* the value of both private and public construction new work in the UK experiencing a record 16.3% fall in 2020.
* construction new orders fell by 11.9% in 2020 reaching its lowest level recorded since 2013, significantly more than the decrease in 2009 after the financial crash.

The pandemic brought with it new ways of working transforming those who had previously been highly suspicious that working from home equated to, watching daytime TV and walking the dog, to be strongly in favour and would seek to embed remote working on future projects.

Since the early days of *New Aspects of Quantity Surveying Practice,* emphasis has been placed on the necessity for professionals in the construction industry to work in an ethical and transparent way. It therefore came as a surprise to some that the RICS was exposed in 2019 for running the institution like the personal fiefdom of the CEO Sean Tompkins by concealing some very unconventional accounting procedures and dismissing the whistle blowers who exposed the affair.

Without doubt sustainability will dominate the ways in which the construction industry and its professionals conduct themselves in the future. RICS Futures 2020 concluded that the profession had a central role to play in shaping solutions to this and other challenges, provided 'we are brave in our thinking and collaborative in our approach.'

The construction industry is one of the least automated industries relying on manual-intensive labour as a primary source of productivity. This edition contains a chapter on digital technology, which without doubt is going to be a major disruptor to conventional working practices.

Duncan Cartlidge
www.duncancartlidgeonline.com

Acknowledgements

My thanks go to the following for contributing to this book:

Siobhan Morrison PgC LTHE, MRICS, FHEA
Siobhan Morrison is programme leader for the BSc Quantity Surveying Programme within the Department of Construction and Surveying at Glasgow Caledonian University. Following a successful career in private practice, Siobhan entered academia in 2017. A contributor to academic conferences, and prestigious journal publications in the field of construction management and economics, Siobhan has also achieved her postgraduate certificate in learning and teaching in higher education where she attained Fellow of the Higher Education Academy and has been involved in research projects with ZeroWasteScotland on the True Cost of Construction Waste. Her primary interest is sharing her passion for construction with the next generation and her research focus lies in the law/contract and procurement of construction projects, as well as how technology can be harnessed to improve the built environment sector and aid the sustainability/environmental agenda.

1 The journey so far

Introduction

The quantity surveyor has been an integral part of the UK construction industry for over 170 years. The golden age for quantity surveyors was the post war period between 1950 and 1980, when bills of quantities were king and the preferred basis for tender documentation and the Royal Institution of Chartered Surveyors (RICS) scale of fees were generous and unchallenged. As described in the following chapter, this situation was due to change during the latter stages of the twentieth century and early part of the twenty-first century as a client-led crusade from both the public and private sectors for, value for money, reliable project completion times and less traditional procurement strategies put new demands on the construction industry and its professions. More recently, a series of government reports called for improved performance and set demanding targets for UK construction.

Construction is very important to the European economy. It provides 18 million direct jobs and contributes to about 9% of the EU's GDP. Despite the poor performance figures (see Figure 1.1), it has been forecast that construction will be the engine of global economic growth in the decade to 2030, with output expected to be 35% higher than in the ten years to 2020. *(Future of Construction, Oxford Economics and Marsh McLennan).* Several factors including pent-up household savings, pandemic stimulus programmes and population growth will combine to spur average annual growth of 3.6% between now and 2030. A quantity surveyor entering the industry today can expect a 50-year plus career, however, the changes encountered over this period will mean their entry training is unlikely to provide all the skills needed to see them through to their retirement. The pace of change has never been so rapid and so relentless and the industry and the quantity surveyor must embrace the changes making the need for life-long learning even more vital. This expansion is expected to exceed growth in both manufacturing and services with the biggest challenges for quantity surveying come from;

DOI: 10.1201/9781003293453-1

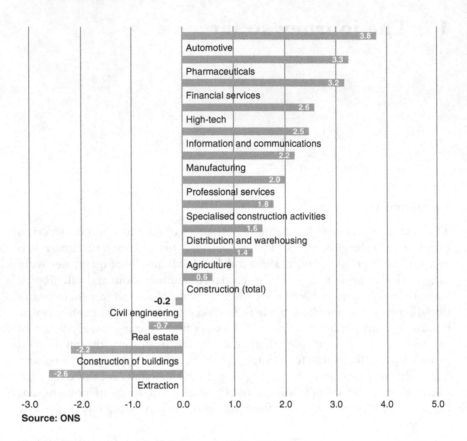

Figure 1.1 UK sector productivity growth 1995–2020.

Data and technology

The construction industry generates large volumes of data but much of it is currently unused. There is scope to make construction projects more efficient, greener and buildings safer, by harnessing the power of data as discussed in Chapter 4 and

Sustainability

Construction is a major polluter and rapid urban growth in the coming decades will add to the problem. The UN predicts that two-thirds of the world's population will be living in cities by 2050 – approximately 6.5 billion people. Two recent reports the first, RICS Futures (2020), outlines the challenges facing built environment professionals while the second, UK Government's Construction Strategy 2025, quantifies the challenge by calling for;

- 33% reduction in initial and whole life costs
- 50% reduction is green-house gases in the built environment.

Twenty-five per cent of the UK's total greenhouse gas emissions are attributable to the built environment. Greenhouse gases are emitted at every stage of the construction and use cycle, from the manufacture of materials through construction and maintenance to eventual demolition. Emissions from the built environment must be reduced if the UK is to meet the net zero target by 2050. More pressingly, the UK's Sixth Carbon Budget requires carbon emissions to be reduced by 78% by 2035, compared to 1990 levels. At COP26, the UK Government committed the UK to achieving a 68% reduction in the UK's carbon emissions by 2030, compared to 1990 levels, as discussed in Chapter 2.

Ethics

In an increasingly transparent world, where even small misdemeanours are forensically scrutinised, it is vital for the construction industry and quantity surveyors to been seen to be conducting themselves ethically and honestly. How unfortunate therefore that the RICS, the body responsible for ensuring that its' member act in accordance with the rules of conduct were themselves exposed as acting unethically in 2019, as will be discussed in Chapter 5.

The Next Normal in Construction, (2020) McKinsey and Company, written at the height of the global pandemic, also forecast changes on the horizon for construction, worldwide. These changes or 'emerging disruptions' were described as:

- Industrialisation and standardisation; new production technology enabling a shift toward off-site production
- New material technology; new, lighter weight materials enabling improved logistics
- Digitalisation of processes and products and shift toward more data driven decision making which will impact:

  - *Operations* – smart buildings and infrastructure
  - *Design* – Building Information Modelling (BIM) objects
  - *Construction and production* – BIM, project management
  - *Channels* – digital sales channels and distribution/logistics

- New entrants on both the supply and demand side – disrupting current business.

If the above wasn't enough, the industry and professions has also had to cope with the impact of two major political and economic events, BREXIT and COVID.

BREXIT/COVID

The debate on the possible impact of BREXIT on construction would seem to have a mixed response. Let's be honest, access to the single market and in particular, public-sector procurement was a non-event for the UK construction industry. I can remember quite clearly the excitement pre-1992 about the opportunities that lay ahead for UK construction and professionals on the European streets paved with gold. Senior partners imagined themselves travelling around Europe to visit their many sites, life would be one big holiday. However, the reality was somewhat different as it quickly became apparent that the European single market was the European closed shop.

As the UK prepared to leave the single market, the amount of true EU cross-border procurement in all sectors was estimated at being somewhere between 6% and 9%. Indirect cross-border procurement, where a firm opens a branch in another member state, is slightly higher. Why was this? Because the EU public procurement regime is not fit for purpose. The EU procurement juggernaut is administered by an under resourced directorate in Brussels and that turns a blind eye to transgressions of the rules. Most EU states are determined to keep large public works contracts within their own borders for political reasons. If there was transgression, then the penalties were non-existent, with whistle blowers being bought off. Works contracts are advertised in the OJEU in English, but the contract documentation is usually in a State's own language, which necessitated a lengthy and more importantly, a costly translation process before the bidding process could even begin. The main obstacles to bidding cross-border have been found to be (Figure 1.2):

1   High competition from national bidders − 40%
2   Perceived preference among contracting authorities for local bidders − 39%
3   Unfamiliar legal context or formal requirements, e.g., contract, labour law, certificates to provide such as special permits for offering services abroad etc., leading to market entry barriers in the awarding country − 32%
4   Additional costs due to geographic distance − 30%
5   Language barriers − 23%.

Furthermore, almost half of small and medium sized enterprises (SMEs) (46%) also reported that identifying sources to access information on cross-border public procurement was a challenge (Figure 1.2).

COVID

As it was for most sectors, 2020 was a highly unusual year for the construction industry but despite recent challenges it has recovered well from the large falls experienced at the start of the pandemic. A lot of work was paused on building sites during the early weeks of the first lockdown in April and May 2020,

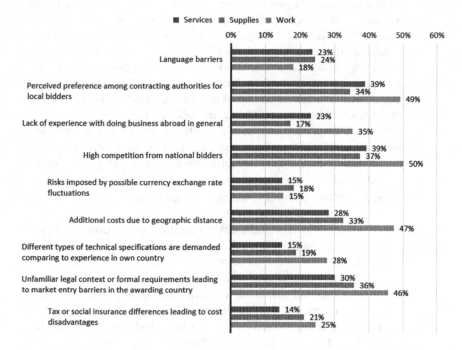

Figure 1.2 Measurement of impact of cross-border penetration in public procurement 2017.

Source: EU Directorate-General for Internal Market.

although activity increased quite rapidly in the second half of the year when the economy reopened, as the industry adapted to new ways of working.

- The value of new UK construction work experienced a record 16.3% fall in 2020 to £99,651 million, in current prices, after the record high of £119,087 million in 2019 as the impact of Coronavirus (COVID-19) pandemic took hold; there were similar percentage reductions in both private and public new work
- New construction orders fell by 11.9% in 2020 to £55,631 million, reaching its lowest level recorded since 2013, driven down by the private housing, private commercial (mainly entertainment) and public other new work sectors; the only sectors with positive growth were infrastructure and public new housing, with the former seeing most notable growth in electricity infrastructure and roads
- Overall, total new work fell £19,436 million in 2020 as the impact of the Coronavirus (COVID-19) pandemic took hold. This is significantly more than the £14,082 million decrease in 2009 after the financial crash

- Private housing and private commercial new work contributed three-quarters of the overall decrease in 2020. Both sectors had exceptionally large falls in April and May 2020 as building sites closed because of official guidance on restrictions of movement in Great Britain. The trend for house renovations and extensions, given people were spending more time in their homes than usual, helped boost activity along with businesses undertaking refurbishments in premises to implement social distancing measures effectively
- Private infrastructure was the only sector to see an annual increase in new work (£1,571 million), recording its highest value in the last decade. In contrast, public infrastructure new work fell £2,306 million in 2020.

Source: ONS Construction statistics, Great Britain: 2020

Overall, the impact of BREXIT and COVID on the construction industry has been significant. In a survey conducted by Turner and Townsend in 2021, it was concluded that,

- 78% of respondents said that rising costs of construction had a significant or high impact on the delivery of construction projects
- 67% of respondents said skilled labour shortages had a significant or high impact on the delivery of construction projects
- 33% of respondents said that excessive lead times had a significant or high on impact the delivery of construction projects.

However, by 2022, the speed of the bounce back by the construction industry appeared to dispel the some of the more pessimistic forecasts (Figure 1.3).

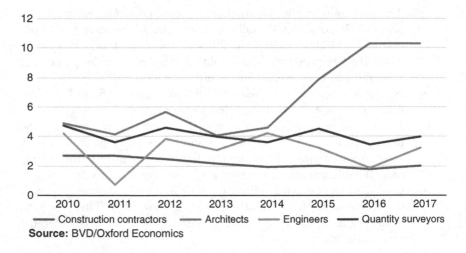

Source: BVD/Oxford Economics

Figure 1.3 Median construction profit margins percentages 2010–2017.

A shift to blended working

In what is described as a snapshot on the impact of COVID on construction, The University of Loughborough *(COVID-19 and construction: Early lessons for a new normal? Loughborough University August 2020)* interviewed a range of professionals about their COVID experiences, including construction directors, construction managers, sub-contractors and health and safety professionals. The increase in working from home for those in suitable roles was identified as a generally positive trend, with potential for long-term use, but with some potential downsides. The positives were thought to be:

- Improved productivity and reduced distraction were reported (these factors were also reported by those who remained working in the offices, which now have fewer occupants)
- Reduced travelling, improved work-home balance
- Reduced costs (for individuals and projects).

However, the negative aspects were identified as:

- Social isolation and poorer wellbeing/mental health for some
- Risk of over-working for some
- Lost benefits of 'grapevine' interactions.

Several interviewees who had previously been highly suspicious that working from home equated to watching daytime TV were now strongly in favour and would seek to embed its use on future projects provided it was actively managed, for example:

- Suitable workspace and equipment will be essential, for example, not working on laptops at the kitchen table
- Regular 'site days' will be essential to maintain relationships, ensure employees stay up to date and maintain team cohesion to avoid any 'us and them' mentality
- Clear expectations are needed regarding expectations and working hours; hours might be variable and flexible but must still be constrained.

The biggest change in technology use was the increase in remote meetings. This is another area where the changes made in response to COVID-19 may well become permanent. Many commented on how effective remote meetings had been and were amazed at how they and their colleagues had adjusted, even those with relatively low IT literacy. In addition to reducing time needed for travel, remote meetings were seen as being more efficient (e.g., one hour rather than two or three), with less time wasted getting cups of tea or having toilet breaks, and less chit chat. The fact that there were fewer meetings overall was also identified as a benefit – many meetings now just do not happen, involve far fewer people or have been replaced by site-based discussions.

Many leading quantity surveying firms have been the industry's keenest advocates for new ways of working, with some firms stating they want to cut down the number of desks by as much as 40% with staff expected to come into the office two to three days a week. One large quantity surveying practice added: '*It's about how we develop meeting space, communal space rather than the factory of 9–5. That thinking has gone*'. In a major study by consultancy Timewise (https:// timewise.co.uk/) in 2022 involving four firms – Skanska, BAM Construct, BAM Nuttall and Willmott Dixon – all adopted new models of working on a range of sites across the country. The goal was to identify if it was possible to achieve project timelines without budgets or deadlines being affected, across a range of sites and projects. The study took place in a range of locations, from a HS2 site in London through to a substation build near Weston-Super-Mare and with teams of between 14 and 120 workers. None of the firms reported negative impacts on budgets or timeframes and some suggested there may have been productivity gains with the proportion of workers who said their working hours gave them enough time to look after their own health and wellbeing increased from 48% to 84%. A year on, all four contractors have chosen to continue with their flexible practices, and all have reported a drop in illness rates among their employees, bucking the post-pandemic industry trend. At the start of the project, nearly half of all participants felt guilty if they started later or finished earlier than others onsite. This portion decreased to a third. Trust in colleagues working remotely increased with the number querying whether colleagues working away from site were working as hard as those on jobs falling from 48% to 33%.

Response to change

How has the industry and quantity surveyors responded to the challenge of change? In the first instance, the changes that took place in the UK construction industry and quantity surveying practice during the latter half of the twentieth and the first part of the twenty-first century. This period sets the scene for the remaining chapters, which go on to describe how quantity surveyors are adapting to new and emerging markets and responding to client-led demands for added value.

The end of the gravy chain

Many of small and medium-size practices that flourished in the period 1960–1990 have now disappeared, victims of mergers and acquisitions of the ever-growing mega practices and consequently, the surveying profession has polarised into two groups:

- The large multidisciplinary practices capable of matching the problem-solving capabilities of the large accountancy-based consulting firms

• Small practices that can offer a fast response from a low-cost base for clients, as well as providing services to their big brother practices.

The construction industry is no stranger to fluctuations in workload, the most recent being the downturn in construction orders following the world financial crisis in 2008 after the collapse of some major American financial institutions left the world on the brink of a 1920s' style depression. The housing sector was particularly badly affected and the downturn resulted in an 8.2% (67,000) fall in construction-related jobs, the largest of any major industrial sector during this period (ONS). By 2016, the industry was showing signs of a strong recovery with skills shortages being reported at all levels, but not before a cull of many established construction firms and many quantity surveying practices running for cover into the arms of large multinational organisations.

As dramatic and concerning as the 2008–2012 downturn was, it was the period between 1990 and 1995 that will be remembered, as an eminent politician once remarked, as 'the mother of all recessions'. Certainly, from the perspective of the UK construction industry, this recessionary phase was the catalyst for many of the changes in working practices and attitudes that have been inherited by those who survived this period and continue to work in the industry. As described in the following chapters, some of the pressures for change in the UK construction industry and its professions – including quantity surveying – have their origins in history, while others are the product of the rapid transformation in business practices that took place during the last decades of the twentieth century and continue today. Perhaps the lessons learnt during this period enabled the industry to weather the 2009 financial storm more easily than otherwise would have been then case. This chapter will therefore examine the background and causes of these changes, and then continue to analyse the consequences and effects on contemporary quantity surveying practice.

Historical overview

1990 was a watershed for the UK construction industry and its associated professions.

As illustrated in Figure 1.4, by 1990, a 'heady brew of change' was being concocted on fires fuelled by the recession that was starting to have an impact on the UK construction industry. The main ingredients of this brew, in no particular order, were:

• The traditional UK hierarchical structure that manifested itself in a litigious, fragmented industry, where contractors and their supply chains were excluded from most of the design decisions
• Changing patterns of workload due to the introduction of fee competition and compulsory competitive tendering

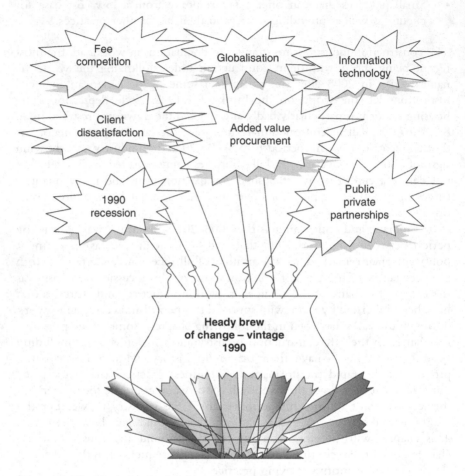

Fee competition

Globalisation

Information technology

Client dissatisfaction

Added value procurement

1990 recession

Public private partnerships

Heady brew of change – vintage 1990

Figure 1.4 The Heady brew of change.

- Widespread client dissatisfaction with the finished product
- The emergence of privatisation and public private partnerships
- The pervasive growth of information technology
- The globalisation of markets and clients.

A fragmented and litigious industry

Boom and bust in the UK construction industry has been and will continue to be a fact of life (see Figure 1.5), and much of the industry, including quantity surveyors, have learned to survive and prosper quite successfully in this climate. While some are large players, at least 99.9% of firms are SMEs and of those, some 83% employ no more than one person. The industry tends to rely on a high degree of sub-contracting and a high proportion of self-employment

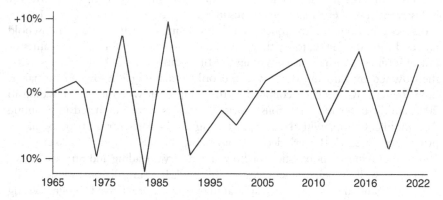

Figure 1.5 Construction output – percentage change 1965–2022.

Source: Dept. for Business, Innovation and Skills.

with over 40% of construction contracting being self-employed. A positive outcome of this is that it makes the industry highly flexible and responsive to changes in the market demand.

The rules were simple: in the good times, a quantity surveyor earned fee income as set out in the Royal Institution of Chartered Surveyors' Scale of Fees for the preparation of, say, bills of quantities, and then in the lean times endless months or even years would be devoted to performing countless tedious re-measurements of the same work – once more for a fee. Contractors and subcontractors won work, albeit with very small profit margins, during the good times, and then when work was less plentiful, they would turn their attentions to the business of the preparation of claims for extra payments for the inevitable delays and disruptions to the works. The standard forms of contract used by the industry, although heavily criticised by many, provided the impetus (if impetus were needed) to continue operating in this way. Everyone, including most clients, appeared to be quite happy with the system, although in practice, the UK construction industry was in many ways letting its clients down by producing buildings and other projects that were, in a high percentage of cases, over budget, over time and littered with defects. Time was running out on this system, and by 1990, the hands of the clock were at five minutes before midnight. A survey conducted in the mid-1990s by Property Week, a leading property magazine, among private sector clients who regularly commissioned new buildings or refurbished existing properties provided a snapshot of the UK construction industry at that time. In response to the question 'Do projects finish on budget?'; 30% of those questioned replied that it was quite usual for projects to exceed the original budget.

In response to the question 'Do projects finish on time?', once again more over 30% of those questioned replied that it was common for projects to overrun their planned completion by one or two months. Parallels between the construction industry ethos at the time of this survey and the UK car industry of the 1960s make an interesting comparison. Austin, Morris, Jaguar, Rolls Royce, Lotus and marques such as the Mini and MG were all household names during the 1960s; today they are all either owned by foreign companies or out of business. At the time of writing the first edition of this book, Rover/MG, then owned by Phoenix (UK), was the only remaining UK-owned carmaker, now Rover/MG has been confined to the scrap heap when it ceased trading in 2005 amidst bitter recriminations and its key assets were purchased by Nanjing Automobile Group, with Nanjing restarting MG sports car and sports saloon production in 2007. Rover's decline from being the UK's largest carmaker in the 1960s is a living demonstration of how a country's leading industry can deteriorate, as well as being a stark lesson to the UK construction industry. The reasons behind the collapse of car manufacturing were flawed design, wrong market positioning, unreliability and poor build quality but importantly to this can be added lack of investment in new technology, and a failure to move with the times and produce what the market; i.e., the end users, demanded. Therefore, when the first Datsun cars began to arrive from Japan in the 1970s and were an immediate success, it was no surprise to anyone except the UK car industry. The British car buyer, after overcoming initial reservations about purchasing a foreign car, discovered a product that had nearly 100% reliability, contained many features as standard that were extras on British-built cars, were delivered on time, and benefitted from long warranties. Instead of producing what they perceived to be the requirements of the British car buyer, Datsun had researched and listened to the needs of the market, seen the failings of the home manufacturers, and then produced a car to meet them. Not only had the Japanese car industry researched the market fully; it had also invested in plant and machinery to increase build quality and reduce defects in their cars. In addition, the entire manufacturing process was analysed and a lean supply chain established to ensure the maximum economies of production. The scale of the improvements achieved in the car industry is impressive, with the time from completed design to launch reduced from 40 to 15 months, and the supplier defects to five parts per million. So why by 1990 was the UK construction industry staring into the same abyss that the carmakers had faced 30 years earlier? To appreciate the situation that existed in the UK construction industry in the pre-1990 period, it is necessary to examine the working practices of the UK construction industry, including the role of the contractor and the professions at this time. First, we will look back at recent history, and in particular at the events that took place in Europe in the first part of the nineteenth century and helped to shape UK practice (Goodall, 2000).

The UK construction industry – a brief history

Prior to the Napoleonic Wars, Britain, in common with its continental neighbours, had a construction industry based on separate trades. This system still exists in France as 'lots sépare', and variations of it can be found throughout Europe, including in Germany. The system works like this: instead of the multi-traded main contractor that operates in the UK, each trade is tendered for and subsequently engaged separately under the co-ordination of a project manager, or 'pilote'. In France, smaller contractors usually specialise in one or two trades, and it is not uncommon to find a long list of contractors on the site board of a construction project. The Napoleonic Wars, however, brought change and nowhere more so than in Britain – the only large European state that Napoleon failed to cross or occupy. Paradoxically, the lasting effect the Napoleonic Wars had on the British construction industry was more profound than on any other national construction industry in Europe. Whilst it is true that no military action took place on British soil, nonetheless the government of the day was obliged to construct barracks to house the huge garrisons of soldiers that were then being transported across the English Channel. As the need for the army barracks was so urgent and the time to prepare drawings, specifications, etc. was so short, the contracts were let on a 'settlement by fair valuation based on measurement after completion of the works'. This meant that constructors were given the opportunity and encouragement to innovate and to problem solve – something that was progressively withdrawn from them in the years to come. The same need for haste, coupled with the sheer magnitude of the individual projects, led to many contracts being let to a single builder or group of tradesmen 'contracting in gross', and the general contractor was born. When peace was made, the Office of Works and Public Buildings, which had been increasingly concerned with the high cost of measurement and fair value procurement, in particular the construction of Buckingham Palace and Windsor Castle, decided enough was enough. In 1828, separate trades contracting was discontinued for public works in England in favour of contracting in gross. The following years saw contracting in gross (general contracting) rise to dominate, and with this development, the role of the builder as an innovator, problem solver and design team member was stifled to the point where contractors operating in the UK system were reduced to simple executors of the works and instructions (although in Scotland the separate trades system survived until the early 1970s). However, history had another twist, for in 1834 architects decided that they wished to divorce themselves from surveyors and establish the Royal Institute of British Architects (RIBA), exclusively for architects. The grounds for this great schism were that architects wished to distance themselves from surveyors and their perceived '*obnoxious commercial interest in construction*'. The top-down system that characterises so much of British society was stamped on the construction industry. As with the death of separate trades contracting, the establishment of the RIBA ensured that the UK contractor was once again discouraged from using innovation. The events of 1834 were also responsible for the birth of another UK phenomenon, the quantity surveyor, and for another unique feature of the UK construction industry – post construction liability.

The ability of a contractor to re-engineer a scheme/concept design in order to produce maximum buildability is a great competitive advantage, particularly on the international scene. A system of project insurance that is already widely available on the Continent is starting to make an appearance in the UK; adopting this, the design and execution teams can safely circumvent their professional indemnity insurance and operate as partners under the protective umbrella of a single policy of insurance, thereby allowing the interface of designers and contractors. However, back to history. For the next 150 or so years, the UK construction industry continued to develop along the lines outlined earlier, and consequently by the third quarter of the twentieth century, the industry was characterised by powerful professions carrying out work on comparatively generous fee scales, contractors devoid of the capability to analyse and refine design solutions, forms of contract that made the industry one of the most litigious in Europe, and procurement systems based upon competition and selection by lowest price and not value for money. Some within the industry had serious concerns about procurement routes and documentation, the forms of contract in use leading to excess costs, suboptimal building quality and time delays, and the adversarial and conflict-ridden relationships between the various parties. A series of government-sponsored reports (Simon, 1944; Emmerson, 1962; Banwell, 1964) attempted to stimulate debate about construction industry practice, but with little effect.

The recession of 2008–2012 triggered another round of quantity surveying practices mergers and acquisitions, this time somewhat more radical. One such take-over in 2010 saw the global engineering giant AECOM acquire Davis Langdon for £204 million; a practice that was established in 1919 by Horace W. Langdon. AECOM has 45,000 employees internationally. In the period 1945–1980, Davis, Langdon and Everest, later to be Davis Langdon, was a flagship-chartered quantity surveying practice, known throughout the world. Global organisations, such as AECOM, bring new / different values and three years after the acquisition Davis Langdon had their name changed to 'AECOM Building and Places'. Horace Langdon would be turning in his grave. Such was the status of Davis Langdon and Everest that in 1991 they were tasked by the RICS to prepare the report QS 2000, a vision of what quantity surveying practice would look like at the turn of the millennium.

The rise and fall of Davis Langdon are chronicled here:

- **1919** Horace W Langdon sets up practice in Holborn, London. He is joined two years later by Tom Every to form Langdon and Every
- **1931** Owen Davis sets up a practice nearby
- **1944** A close working relationship developed between Davis, Belfield and Bobbie Everest, a quantity surveyor and descendent of George Everest, Surveyor General of India from 1830–1843 and the first man to measure the height of Mount Everest, which consequently bore his name and in 1944 they formed the partnership of Davis, Belfield & Everest (DBE)
- **1948–1988** Langdon and Every and DBE expanded during the post-war construction programme and established offices in the UK, the Middle East and Asia-Pacific
- **1988** The two firms merge to form Davis Langdon & Everest, the biggest merger of QS firms to date

- **2004** Davis Langdon & Everest becomes Davis Langdon LLP
- **2008** Davis Langdon fail to respond to the rapid downturn in the market and rather like the Titanic continued full steam ahead – see later comments in this chapter on change management
- **2010** AECOM acquires Davis Langdon's European, Middle East, Australasian, African and US operations for £204m. Partners in the Asian arm Davis Langdon & Seah reject the takeover. This part of the firm being acquired by Arcadis, a Dutch based design, engineering and management consulting company in 2012.

The take-over was not without its problems and subsequently 40% of the long-serving equity partners left the practice in response to the new structure and direction. Rob Smith the then senior partner at Davis Langdon LLP, now retired, wrote the Foreword to the second edition of *New Aspects of Quantity Surveying Practice* in 2006 when the practice's future looked bright and shortly after revealed in the trade press that his earnings for 2005 were £507,000.00 (Building, Ed 44, 2005).

The increase in the number of take-overs and acquisitions has seen once large and influentially practices such as Davis Langdon subsumed into large multi-sectorial organisations. Some quantity surveyors are being encouraged by their new corporate owners to refer to themselves as cost managers/cost engineers. In addition to AECOM, conglomerates such as Serco and CAPITA have swallowed several established quantity surveying practices without perhaps really understanding what they were getting for their money. Many within quantity surveying circles see this trend in take-overs as a change from the tradition ethos of providing client services to maximising profit for shareholders through control of headcount.

In 2016, yet another take-over of a traditional quantity surveying practice Cyril Sweett by Middle East owned firm Currie & Brown after Cyril Sweett's bankers withdrew their financial support. See later comments in Chapter 5. The price paid for Sweetts was reported to be £29 million and the takeover, completed by September 2016 de-listed Sweett from the AIM arm of the London Stock Exchange after nine years. The takeover, which created a combined company of over 2,000 employees, ended nearly 90 years of independence at Sweett. Currie and Brown moved swiftly to end the Sweett brand as they announced '*you can't have two names in the same marketplace*'. Currie and Brown stated that cost savings were to be made, principally through de-listing and cutbacks in property, back office and shared services. By a strange coincidence, Francis Ives, the chairman of Cyril Sweett from 1991–2010 wrote the forward to the third edition of *New Aspects of Quantity Surveying Practice*; it's proving to be quite a poisoned chalice!

During the preparation of the various editions of *New Aspects of Quantity Surveying Practice* the constant stream of government reports, that regaled the construction industry for being inefficient, dysfunctional and providing poor value for money, just kept on coming.

Top 16 industry reports since Egan's Rethinking Construction in 1998 are as follows:

- Achieving Excellence, Office of Government Commerce, 1999
- Modernising Construction, National Audit Office (NAO), 2001
- Accelerating Change, Strategic Forum for Construction, 2002
- Improving Public Services Through Better Construction, NAO, 2005
- Be Valuable, Constructing Excellence, 2005
- Callcutt: Review of Housebuilding Delivery, John Callcutt, 2007
- Construction Commitments, Strategic Forum for Construction, 2008
- The Strategy for Sustainable Construction 2008
- Government/Strategic Forum, 2008
- Construction Matters, Business and Enterprise Select Committee, 2008
- Equal Partners, Business Vantage and Construction Clients' Group, 2008
- Never waste a good crisis – The Wolstenholme Report, Constructing Excellence, 2009
- Government Construction Strategy 2011, Cabinet Office
- Construction 2025, Construction Industrial Advisory Council, 2013
- The National Infrastructure Strategy, HM Treasury, 2020
- The Construction Industry Playbook, The Cabinet Office, 2020.

It's all well and good producing reports but does the construction industry take any notice of them. The four key projected processes, according to Rethinking Construction, needed to achieve change were identified as:

1   *Partnering in the supply chain* – development of long-term relationships based on continuous improvement within the supply chain
2   *Components and parts* – a sustained programme of improvement for the production and the delivery of components with an emphasis on modern methods of construction (MMC)
3   *Focus on the end product* – integration and focusing of the construction process on meeting the needs of the end user
4   *Construction process* – the elimination of waste.

In addition, annual targets capable of being achieved in improving the performance of construction projects were collectively identified as follows:

1   To reduce capital costs by 10%
2   To reduce construction time by 10%
3   To reduce defects by 20%
4   To reduce accidents by 20%
5   To increase the predictability of projected cost and time estimates by 10%
6   To increase productivity by 10%
7   To increase turnover and profits by 10%.

Can there be any other industry that has generated so many reports?

Government construction strategy 2011

The Government's Plan for Growth, published alongside Budget 2011, highlighted the critical importance of an efficient construction industry to the UK economy. The report focused on public sector work; some 40% of construction work comes from the public sector, with central Government being the industry's biggest customer. It was claimed, yet again in the study, that the UK does not get full value from public sector construction, and that it has failed to exploit the potential for public procurement of construction and infrastructure projects to drive growth. This strategy called for a profound change in the relationship between public authorities and the construction industry. The stated aspiration is to be more strategic and less transactional to ensure the Government consistently gets a good deal and the country gets the social and economic infrastructure it needs for the long-term. There is a strong commitment to introduce wide use of Modern Methods of Construction thereby transitioning the construction process from one of building to one of assembly.

A detailed programme of measures called for:

- Replacement of adversarial cultures with collaborative ones
- Cost reduction and innovation within the supply chain to maintain market position rather than innovation that is focussed on the bidding process, with a view to establishing a bargaining position for the future.

The proposed model, for public sector construction procurement in the UK as detailed in *The Construction Playbook* (see Chapter 3) is one in which:

- Clients issue a brief that concentrates on required performance and outcome, designers and constructors work together to develop an integrated solution that best meets the required outcome
- Contractors engage key members of their supply chain in the design process where their contribution creates value
- Value for money and competitive tension are maintained by effective price benchmarking and cost targeting
- By knowing what projects should cost, rather than through lump sum tenders based on inadequate documentation, supply chains are, where the programme is suited, engaged on a serial order basis of sufficient scale and duration to incentivise research and innovation around a standardised (or mass customised) product
- Industry is provided with sufficient visibility of the forward programme to make informed choices (at its own risk) about where to invest in products, services, technology and skills
- There is an alignment of interest between those who design and construct a facility and those who subsequently occupy and manage it.

Approaches to help achieve this reduction were outlined as follows:

* The use of BIM. Government required fully collaborative 3D BIM (with all project and asset information, documentation and data being electronic) as a minimum by 2016
* The introduction of Soft Landings and Government Soft Landings which aim to ease the transition from completion to commissioning with the alignment of design/construction with operation and asset management
* Post-handover defects are a regular feature of construction projects, leading to the cost of remediation (and frequently the higher cost of resolving disputes). Even when there are no latent defects, it is still rare to find that a built asset performs exactly in accordance with its design criteria (and particularly in terms of energy efficiency, for example)
* Integration of the design and construction of an asset with the operation phase should lead to improved asset performance. This has been demonstrated in projects which have integrated design and construction with whole-life operation. The same alignment can be created by requiring those who design and construct buildings to prove their operational performance for a period of say three to five years. Proposals for this will be developed with the Government Property Unit to ensure alignment with subsequent arrangements for facilities management.

BIM will be discussed more fully later in this chapter as well as in Chapter 4.

Construction 2025

In July 2013 another Government report, Construction 2025, hit the book-stands, this time targeted at both the public and private sectors and outlined a vision for the construction industry by 2025. Even more targets were set as follows:

* 33% reduction in the initial cost of construction and whole life costs
* 50% reduction in overall time from inception to delivery
* 50% lower green-house emissions
* 50% improvements in exports.

It was not just the UK construction industry that was obsessed with navel-gazing quantity surveyors had also been busy penning numerous reports into the future prospects for their profession. The most notable of which were: The Future Role of the Chartered Quantity Surveyor (1983), Quantity Surveying 2000: The Future Role of the Chartered Quantity Surveyor (1991) and the Challenge for Change: QS Think Tank (1998), Our Changing World: Let's Be Ready (2015), Future of the Profession (2019), RICS Futures (2020) all produced either directly by or on behalf of The Royal Institution of Chartered Surveyors. The 1971 report, The Future Role of the Quantity Surveyor paints a picture of a world where the quantity surveyor was primarily a producer of bills of quantities; indeed, the report concluded that the distinct competence of the quantity surveyor of the 1970s was measurement – a view, it should be added, still shared by many today. In addition, competitive single stage tendering was the norm, as was the practice of receiving most work via the patronage of an architect. It was a profession where design and construct projects were rare, and quantity surveyors were discouraged from forming multidisciplinary practices and encouraged to adhere to the RICS scale of fees. The report observes that clients were becoming more informed, but there was little advice about how quantity surveyors were to meet this challenge. A mere 25 years later, the 1998 report, The Challenge of Change, was drafted in a business climate driven by information technology, where quantities generation is a low-cost activity and the client base is demanding that surveyors demonstrate added value. In particular, SME-sized quantity surveying firms (i.e., fewer than 250 employees EU Definition) were singled out by this latest report to be under particular pressure owing to:

- Competing with large practices' multiple disciplines and greater specialist knowledge base
- Attracting and retaining a high-quality work force
- Achieving a return on the necessary investment in IT
- Competing with the small firms with low overheads.

Interestingly, the Challenge for Change report also predicts that the distinction between contracting and professional service organisations will blur, a quantum leap from the 1960s, when chartered surveyors were forced to resign from the RICS if they worked for contracting organisations.

RICS futures

In 2015, the RICS launched the Futures project with the publication of an insight paper; *RICS Futures, Our Changing World: Let's Be Ready*. Year 2020 saw the publication of the latest update; The Futures Report 2020. The project which reflects global views starts by outlining social and economic trends and the changing business landscape before concentrating on the changing role of the profession. The debate according to the RICS is whether the future industry will demand specialists or generalists. There is strong emphasis on ethical practice as well as a recognition that the profession must embrace new and emerging technologies. The RICS Futures Report 2020 concluded:

- It is critical for our profession to be able to understand and utilise data effectively
- Artificial Intelligence and the Internet of Things are now core parts of our sector. This brings huge potential, but we need to think about the privacy aspects of these new technologies

- The drive for better connectivity will bring opportunities and risks for the profession as we see increasing convergence between the built environment and technology sectors
- Traditional business models are changing and becoming more decentralised, as is the way real estate is owned, traded and managed. It is no longer just about bricks and mortar, but the consumer
- It will become increasingly important to understand the full lifecycle of an asset, not just its value at a fixed point in time. Professionals also need to think about how we value the digital as well as the physical assets we manage
- The traditional office is going through a paradigm shift and workspaces and smarter utilisation of these buildings will bring significant benefit
- Lifelong learning will become increasingly important
- There is a need to understand and manage the impact of urbanisation on the environment, which will be an increasingly urgent requirement of the profession going forward.

As discussed previously in this chapter mergers and acquisitions result in significant change to traditional quantity surveying firms' structures and ownership models. For many professionals including quantity surveyors, this means adapting to more cross-functional teams and new business cultures. As organisations become larger and provide services to a multitude of industries, the trend towards interdisciplinary working and the demand for generalist skills appears to be growing. This it is though could result in greater awareness of the work processes of fellow professionals, such as architects and engineers, in meeting team goals and client outcomes. However, it is recognised that many firms still require, and will continue to demand, specialist technical skill sets such as traditional quantity surveying competencies. For firms that offer such services, staff with strong technical knowledge are essential. Larger, more generalist organisations that do not maintain such knowledge in-house will rely on the expertise of specialist organisations a trend that was highlighted in *Challenge for Change: QS Think Tank* (1998).

While thinking about the future, the average age of the workforce is rising faster than ever: the Chartered Institute of Building reports that the number of workers over 60 is increasing faster, and the set under 30 decreasing faster, than any other set. There is evidence that construction is perceived by many young people as '*being outdoors and getting dirty*' and most suited to '*young people who do not get into college or university*'. The survey also concluded that construction has difficulty appealing to women being characterised as '*Pale, male and stale*'. More must be done to attract new and diverse talent.

Changing patterns of workload

The patterns of workload that quantity surveyors had become familiar with were also due to change. The disruptors came chiefly from two sources:

1   Fee competition and compulsory competitive tendering (CCT)
2   The emergence of a new type of construction client.

Fee competition and compulsory competitive tendering (CCT)

Until the early 1970s, fee competition between professional practices was almost unheard of. All the professional bodies published scales of fees, and competition was vigorously discouraged on the basis that a client engaging an architect, engineer or surveyor should base his or her judgement on the type of service and not on the level of fees. Consequently, all professionals within a specific discipline quoted the same fee. However, things were to change with the election of the Conservative Government in 1979. The new government introduced fee competition into the public sector by way of its compulsory CCT programme, and for the first time professional practices had to compete for work in the same manner as contractors or subcontractors – that is, they would be selected by competition, mainly on the basis of price. The usual procedure was to submit a bid based upon scale of fees minus a percentage. Initially these percentage reductions were a token 5% or 10%, but as work became difficult to find in the early 1980s, practices offered 30% or even 40% reduction on fee scales. It has been suggested that during the 1980s, fee income from some of the more traditional quantity surveying services was cut by 60%. Once introduced there was no going back, and soon the private sector began to demand the same reduction in fee scales; within a few years, the cosy status quo that had existed and enabled private practices to prosper had gone. The Monopolies and Mergers Commission's 1977 report into scales of fees for surveyors' services led the Royal Institution of Chartered Surveyors to revise its byelaws in 1983 to reduce the influence of fee scales to the level of 'providing guidance' – the gravy train had hit the buffers!

Byelaw 24 was altered from:

No member shall with the object of securing instructions or supplanting another member of the surveying profession, knowingly attempt to compete on the basis of fees and commissions

to

... no member shall ... quote a fee for professional services without having received information to enable the member to assess the nature and scope of the services required.

With the introduction of fee competition, the average fee for quantity surveying services (expressed as a percentage of construction cost) over a range of new build projects was reduced to just 1.7%. As a result, professional practices found it increasingly difficult to offer the same range of services and manning levels on such a reduced fee income; they had radically to alter the way they operated or go out of business. However, help was at hand for the hard-pressed practitioner; the difficulties of trying to manage a practice on reduced fee scale income during the latter part of the 1980s were mitigated by a property boom, which was triggered in part by a series of Government-engineered events that combined to unleash a feeding frenzy of property development. In 1988, construction orders peaked at £26.3 billion, and the flames under the heady brew of change were dampened down, albeit only for a few years. The most notable of these events were as follows:

- The so-called Stock Exchange 'Big Bang' of 1986, which had the direct effect of stimulating the demand for high-tech offices
- The deregulation of money markets in the early 1980s, which allowed UK banks for the first time to transfer money freely out of the country, and foreign finance houses and banks to lend freely on the UK market and invest in UK real estate
- The announcement by the Chancellor of the Exchequer, Nigel Lawson, of the abolition of double tax mortgage relief for domestic dwellings in 1987, which triggered an unprecedented demand for residential accommodation; the result was a massive increase in lending to finance this sector, as well as spiralling prices and land values
- Last but by no means least, the relaxation of planning controls, which left the way open for the development of out-of-town shopping centres and business parks.

However, most property development requires credit, and the boom in development during the late 1980s could not have taken place without financial backing. By the time the hard landing came in 1990, many high street banks with a reputation for prudence found themselves dangerously exposed to high-risk real estate projects. During the late 1980s, virtually overnight the banks changed from conservative risk managers to target-driven loan sellers, and by 1990, they found themselves with a total property related debt of £500 billion. The phenomenon was not just confined to the UK. In France, for example, one bank alone, Credit Lyonnais, was left with Euro 10 billion of unsecured loss after property deals on which the bank had lent money collapsed because of over-supply and a lack in demand; only a piece of creative accountancy and state intervention saved the French bank from insolvency. The property market crash in the early 1990s occurred mainly because investors suffered a lack of confidence in the ability of real estate to provide a good return on investment in the short to medium term in the light of high interest rates, even higher mortgage rates, and an inflation rate that doubled within two years. In part, it was also brought about by greed because of the knowledge that property values had historically seldom delivered negative values. As large as these sums seem they pale into insignificance to the debts rung up by banks like The Royal Bank of Scotland in the period 2005–2008, who reported a £28 billion loss in January 2009 and were only saved from insolvency by a government led bailout.

A new type of construction client

Another vital ingredient in the brew of change was the emergence of a new type of construction client. Building and civil engineering works have traditionally been commissioned by either public or private sector clients. The public sector has been a large and important client for the UK construction industry and its professions. Most Government bodies and public authorities would compile lists or 'panels' of approved quantity surveyors and contractors for the construction of hospitals, roads and bridges, social housing and so on, and inclusion on these panels ensured that they received a constant and reliable stream of work. However, during the 1980s the divide between public and private sectors was to blur. The Conservative government of 1979 embarked upon an energetic and extensive campaign of the privatisation of the public sector that culminated in the introduction of the Private Finance Initiative in 1992. Within a comparatively short period, there was a shift from a system dominated by the public sector to one where the private sector was growing in importance. Despite this shift to the private sector, the public sector remains, for the moment, influential; in 2021, for example, it accounted for 38% of the UK civil engineering and construction industry's business, with a government pledge to maintain this level of expenditure. Nevertheless, the privatisation of the traditional public sector resulted in the emergence of major private sector clients such as the British Airports Authority, privatised in 1987, with an appetite for change and innovation. This new breed of client was, as the RICS had predicted in its 1971 report on the future of quantity surveying, becoming more knowledgeable about the construction process, and such clients were not prepared to sit on their hands while the UK construction industry continued to underperform. Clients such as Sir John Egan, who in July 2001 was appointed Chairman of the Strategic Forum of Construction, became major players in the drive for value for money. The poor performance of the construction industry in the private sector has already been examined, however, if anything, performance in the public sector paints an even more depressing picture. This performance was scrutinised by the National Audit Office (NAO) in 2001 in its report Modernising Construction (Auditor General, 2000), which found that many projects were over budget and delivered late. So dire had been the experience of some public sector clients – for example the Ministry of Defence – that new client driven initiatives for procurement, were introduced. In particular, there were a number of high-profile public projects disasters such as the new Scottish Parliament in Edinburgh, let on a management contracting basis which rose in cost from approximately £100 million to £450 million and was delivered in 2004 – two years late and with a total disregard for life cycle costs.

If supply chain communications were polarised and fragmented in the private sector, then those in the public sector were even more so. A series of high-profile cases in the 1970s, in which influential public officials were found to have been guilty of awarding construction contracts to a favoured few in return for bribes, instilled paranoia in the public sector, which led to it distancing itself from contractors, subcontractors and suppliers – in effect from the whole supply chain. At the extreme end of the spectrum, this manifested itself in public sector professionals refusing to accept even a diary, calendar or a modest drink from a contractor in case it was interpreted as an inducement to show bias. In the cause of appearing to be fair, impartial and prudent with public funds, most public contracts were awarded as a result of competition between a long list of contractors on the basis of the lowest price. The 2001 National Audit Office report suggests that the emphasis on selecting the lowest price is a significant contributory factor to the tendency towards adversarial relationships. Attempting to win contracts under the 'lowest price wins' mentality leads firms to price work unrealistically low and then seek to recoup their profit margins through contract variations arising from, for example design changes and claims leading to disputes and litigation. A sentiment reinforced in *The Construction Playbook* some 20 years later. In an attempt to eradicate inefficiencies, the public sector commissioned a number of studies such as The Levene Efficiency Scrutiny in 1995, which recommended that departments in the public sector should:

- Communicate better with contractors to reduce conflict and disputes
- Increase the training that their staff receives in procurement and risk management
- Establish a single point for the construction industry to resolve problems common to a number of departments. The lack of such a management tool was identified as one of the primary contributors to problems with the British Library project.

In June 1997, it was announced that CCT would be replaced with a system of Best Value to introduce, in the words of the then local government minister Hilary Armstrong, *'an efficient, imaginative and realistic system of public sector procurement'*. Legislation was passed in 1999, and from 1 April 2000, it became the statutory duty of the public sector to obtain best value. Best value is a constant theme throughout this book.

The impact of information technology

As measurers and information managers, quantity surveyors have been greatly affected by information technology. Substantial parts of the chapters that follow are devoted to the influence that IT has had and will continue to have, both directly and indirectly, on the quantity surveying profession. However, this opening chapter would not be complete without a brief mention of the contribution of IT to the heady brew of change. There are a wide variety of specialist software packages available to quantity surveyors to help with measurement and estimating. In addition, spreadsheets are now widely used for less automated approaches to taking off. The speed of development of specialist software has been breath-taking. In 1981, the Department of the Environment developed and used a computer-aided bill of quantities pro-duction package called 'Enviro'. This then state-of-the-art system required the quantity surveyor to code each measured item, and on completion the codes were sent to Hastings, on the South Coast of England, where a team of operators would input the codes, with varying degrees of accuracy, into a mainframe computer. After the return of the draft bill of quantities to the measurer for checking, the final document was then printed, which in most cases was four weeks after the last dimensions were taken off!

Those who mourn the demise of traditional methods of bill of quantities production should at least take heart that no longer will the senior partner be able to include those immortal lines in a speech at the annual Christmas office party – 'you know after twenty years of marriage my wife thinks that quantity surveying is all about taking off and working up' – pause for laughter!

As mentioned previously, there had been serious concern both in the industry and in government about the public image of UK Construction plc. The 1990 recession had opened the wounds in the construction industry and shown its vulnerability to market pressures. Between 1990 and 1992, over 3800 construction enterprises became insolvent, taking with them skills that would be badly needed in the future, the pattern was repeated between 2008 and 2012. The professions also suffered a similar haemorrhage of skills as the value of construction output fell by double digit figures year on year. The recession merely highlighted what had been apparent for years: the UK construction industry and its professional advisors had to change. The heady brew of change was now complete but concerns over whether or not the patient realised the seriousness of the situation still gave grounds for concern. The message was clear; the construction industry and quantity surveying must change or, like the dinosaur, be confined to history.

The Latham report

In traditional manner, the UK construction industry turned to a report to try to solve its problems. In 1993, Sir Michael Latham, an academic and politician, was tasked to prepare yet another review, this time of the procurement and contractual arrangements in the United Kingdom construction industry. In July 1994, Constructing the Team (or The Latham Report, as it became known) was published. The aims of the initiative were to reduce conflict and litigation, as well as to improve the industry's productivity and competitiveness. The construction industry held its breath – was this just another Banwell or Simon to be confined, after a respectful period, to gather dust on the shelf? Thankfully not. The UK construction industry was at the time of publication in such a fragile state that the report could not be ignored. That's not to say that it was greeted with open arms by everyone – indeed, the preliminary report Trust and Money, produced in December 1993, provoked profound disagreement in the industry and allied professions.

Latham's report found that the industry required a good dose of medicine, which the author contended should be taken in its entirety if there was to be any hope of a revival in its fortunes. The Latham Report highlighted the following areas as requiring particular attention to assist UK construction industries to become and be seen as internationally competitive:

- Better performance and productivity, to be achieved by using adjudication as the normal method of dispute resolution, the adoption of a modern contract, better training, better tender evaluation, and the revision of post-construction liabilities to be more in line with, say, France or Spain, where all parties and not just the architect are considered to be competent players including the construction team and suppliers, thereby making them liable for non-performance for up to ten years
- The establishment of well-managed and efficient supply chains and partnering agreements
- Standardisation of design and components, and the integration of design, fabrication and assembly to achieve better buildability and functionality
- The development of transparent systems to measure performance and productivity both within an organisation and with competitors
- Teamwork and a belief that every member of the construction team from client to subcontractors should work together to produce a product of which everyone can be justifiably proud.

The Latham Report placed much of the responsibility for change on clients in both public and private sectors. For the construction industry, Latham set the target of a 30% real cost reduction by year 2000, a figure, it is believed, was based on the CRINE (Cost Reduction Initiative for the New Era) review carried out in the oil and gas industries a few years previously (CRINE, 1994). The CRINE review was instigated in 1992, with the direct purpose of identifying methods to reduce the high operational costs in the North Sea oil and gas industry. It involved a group of operators and contractors working together to investigate the cause for such high costs in the industry, and produce recommendations to aid the remedy of such. The leading aim of the initiative was to reduce development and production costs by 30%, this being achieved through recommendations such as; the use of standard equipment, simplifying and clarifying contract language, removing adversarial clauses, rationalisation of regulations, and the improvement of credibility and quality qualifications. It was recommended that the operators and contractors work more closely, pooling information and knowledge, to help drive down the increasing costs of hydrocarbon products and thus indirectly promote partnering and alliancing procurement strategies (see Chapter 3). The CRINE initiative recommendations were accepted by the oil and gas industries, and it is now widely accepted that without the use of partnering/alliancing a great number of new developments in the North Sea would not have been possible. Shell UK Exploration and Production reported that the performance of the partners in the North Fields Unit during the period 1991–1995 resulted in an increase in productivity of 25%, a reduction in overall maintenance costs of 31% in real terms, and a reduction in platform 'down time' of 24%.

Could these dramatic statistics be replicated in the construction industry? 'C' is not only for construction but also for conservative, and many sectors of the construction industry considered 30% to be an unrealistically high and un-reachable target. Nevertheless, certain influential sections of the industry, in-cluding Sir John Egan and BAA, accepted the challenge and went further declaring that 50% or even 60% savings were achievable. It was the start of the client-led crusade for value for money.

The Latham Report spawned several task groups to investigate further the points raised in the main report, and in October 1997, as a direct result of one of these groups, Sir John Egan, a keen advocate of Sir Michael Latham's report and known to be a person convinced of the need for change within the industry, was appointed as head of the Construction Task Force. One of the Task Force's first actions was to visit the Nissan UK car plant in Sunderland to study the com-pany's supply chain management techniques and to determine whether they could be utilised in construction (see Chapter 6). In June 1998, the Task Force published the report Rethinking Construction (DoE, 1998), which was seen as the blueprint for the modernisation of the systems used in the UK construction industry to procure work. As a starting point, Rethinking Construction revealed that in a survey of major UK property clients, many continued to be dissatisfied with both contractors' and consultants' performance. Added to the now familiar concerns about failure to keep within agreed budgets and completion schedules, clients revealed:

- More than a third of them thought that consultants were lacking in providing a speedy and reliable service
- They felt they were not receiving good value for money insofar as construction projects did not meet their functional needs and had high life cycle costs
- They felt that design and construction should be integrated in order to deliver added value.

Frustrated by the lack of change in the construction industry, Egan's last act before moving on from and closing down the Task Force in 2002 was to pen his final report Accelerating Change.

As for quantity surveyors, the 1990s ended with perhaps the unkindness cut of all. The RICS, in its Agenda for Change initiative, replaced its traditional divisions (which included the Quantity Surveying Division) with 16 faculties, not unlike the system operated by Organisme Professionel de Qualification Technique des Economistes et Coordonnateurs de la Construction (OPQTECC), the body responsible for the regulation of the equivalent of the quantity surveyor in France. It seemed to some that the absence of a quantity surveying faculty would result in the marginalisation of the profession; however, the plan was im-plemented in 2000, with the Construction Faculty being identified as the new home for the quantity surveyor within the RICS. This move however was not taken lying down by the profession, disillusioned quantity surveyors threatened the RICS with legal action to reverse the decision and in 2004 the QSi was formed by Roger Knowles as '*the only professional body that caters solely for quantity surveyors*'. In 2023 the QSi appears still to be open for business there is little information about the numbers of disillusioned quantity surveyors that it has attracted. The QSi objective is; *to support and protect the character, status and interest of the profession of Quantity Surveying*. Membership is being offered for around £100 pa.

Beyond the rhetoric

How are the construction industry and the quantity surveyor rising to the challenges outlined in the previous pages? When the much-respected quantity surveyors Arthur J. and Christopher J. Willis penned the foreword to the eighth edition of their famous book *Practice and Procedure for the Quantity Surveyor* in 1979, the world was a far less complicated place. Diversification into new fields for quantity surveyors included heavy engineering, coal mining and 'working abroad'. In the Willis's book, the world of the quantity surveyor was portrayed as a mainly technical back-office operation providing a limited range of services where, in the days before compulsory competitive tendering and fee competition, 'professional services were not sold like cans of beans in a supermarket'. The world of the Willis's was typically organised around the production of bills of quantities and final accounts, with professional offices being divided into pre- and post-contract services. This model was uniformly distributed across small and large practices, the main difference being that the larger practices would tend to get the larger contracts and the smaller practices the smaller contracts. This situation had its advantages, as most qualified quantity surveyors could walk into practically any office and start work immediately; the main distinguishing feature between practices A and B was usually only slight differences in the format of taking-off paper. However, owing to the changes that have taken place not only within the profession and the construction industry but also on the larger world stage (some of which have been outlined in this chapter), the world of the Willis's has, like the British motor car industry, all but disappeared forever.

In the early part of the twenty-first century, the range of activities and sectors where the quantity surveyor is active is becoming more and more diverse. The small practice concentrating on traditional pre- and post-contract services is still alive and healthy. However, at the other end of the spectrum the larger practices are now rebadged as international consulting organisations and would be unrecognisable to the Willis's. The principal differences between these organisations and traditional large quantity surveying practices are generally accepted to be the elevation of client focus and business understanding and the move by quantity surveyors to develop clients' business strategies and deliver added value. As discussed in the following chapters, modern quantity surveying involves working in increasingly specialised and sectorial markets where skills are being developed in areas including strategic advice in the public–private partnerships (PPPs), partnering, value and supply chain management. From a client's perspective, it is not enough to claim that the quantity surveyor and/or the construction industry is delivering a better value service; this must be demonstrated. Certainly, there seems to be a move by the larger clients away from the traditional low-profit, high-risk, confrontational procurement paths as discussed in Chapter 3.

The terms of reference for the Construction Industry Task Force concentrated on the need to improve construction efficiency and to establish best practice. The industry was urged to take a lead from other industries, such as car manufacturing, steel making, food retailing and offshore engineering, as examples of market sectors that had embraced the challenges of rising worldclass standards and invested in and implemented lean production techniques. Rethinking Construction identified five driving forces that needed to be in place to secure improvement in construction and four processes that had to be significantly enhanced, and set seven quantified improvement targets, including annual reductions in construction costs and delivery times of 10% and reductions in building defects of 20%. The report also drew attention to the lack of firm quantitative information with which to evaluate the success or otherwise of construction projects. Such information is essential for two purposes:

1   To demonstrate whether completed projects have achieved the planned improvements in performance
2   To set reliable targets and estimates for future projects based on past performances.

It has been argued that organisations like the Building Cost Information Service have been providing a benchmarking service for many years through its tender based index. Additionally, what is now required is a transparent mechanism to enable clients to determine for themselves which professional practice, contractor, subcontractor, etc. delivers best value.

Measuring performance

Benchmarking is a generic management technique used to compare performance between varieties of strategically important performance criteria. These criteria can exist between different organisations or within a single organisation provided that the task being compared is a similar process. It is an external focus on internal activities functions or operations aimed at achieving continuous improvement. Construction, because of the diversity of its products and processes is one of the last industries to embrace objective performance measurements. There is a consensus among industry experts that one of the principal barriers to promote improvement in construction projects is the lack of appropriate performance measurement and this was referred to also in Chapter two in relation to whole life costs calculations. For continuous improvement to occur, it is necessary to have performance measures which check and monitor performance, to verify changes and the effect of improvement actions, to understand the variability of the process and in general it is necessary to have objective information available in order to make effective decisions. Despite the late entry of benchmarking to construction, this does not diminish the potential benefits that could be derived; however, it gives some indication of the fact that there is still considerable work to be undertaken both to define the areas where benchmarking might be valuable and the methods of measurement. The current benchmarking and KPI programme in the UK construction industry has been headlined as a way to improve underperformance.

The Xerox Corporation in America is considered to be the pioneer of benchmarking. In the late 1970s, Xerox realised that it was on the verge of a crisis when Japanese companies were marketing photocopiers cheaper than it cost Xerox to manufacture a similar product. It is claimed that by benchmarking Xerox against Japanese companies, it was able to improve their market position and the company has used the technique ever since to promote continuous improvement. Yet again another strong advocate of benchmarking is the automotive industry who successfully employed the technique to reduce manufacturing faults. Four types of benchmarking can broadly be defined; Internal, Competitive, Functional and Generic (Lema and Price, 1995). However, Carr and Winch (1998) whilst regarding these categories as important suggest that a more useful distinction in terms of methodology is that of output and process benchmarking.

Through the implementation of performance measures (what to measure) and selection of measuring tools (how to measure), an organisation or a market sector communicates to the outside world and clients the priorities, objectives and values that the organisation or market sector aspires to. Therefore, the selection of appropriate measurement parameters and procedures is very important to the integrity of the system (Figure 1.6).

Key performance indicators

It is true to say that most organisations that participate in the production of Key Performance Indicators (KPIs) for the Construction Best Practice Programme (CBPP) has to date produce benchmarks. Since the later 1990s there has been a widespread Government backed campaign to introduce benchmarking into the construction industry with the use of so called Key Performance Indicators (KPIs). The objectives of the benchmarking as defined by *Best Practice in Benchmarking, The Cabinet Office* (2018) and provides an indication of position relative to what is considered optimum practice and hence indicate a goal to be obtained. While useful for getting a general idea of areas requiring performance improvement, they provide no indication of the mechanisms by which increased performance may be brought about. Basically, it tells us that we are underperforming but it does not give us the basis for the underperformance. The production of KPIs which has been the focus of construction industry initiatives to date therefore has been

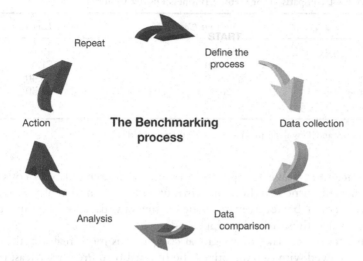

Figure 1.6 Benchmarking process.

Source: RICS Black Book Cost analysis and benchmarking RICS guidance note (2013).

concentrated on output benchmarks. Benchmarking projects have tended to re-main as strategic goals at the level of senior management.

The UK system compared

As discussed earlier in this chapter, there has always been a suspicion, especially by government, that the UK construction industry is inefficient and delivers poor value for money. A report by Turner and Townsend in 2022 (see Table 1.1) seems to suggest that the out-turn cost of construction is more in the UK than in other countries of the world, despite the fact that the UK has quantity surveyors. For many years similar comparative studies have been produced purporting to show how expensive the UK is for construction most of which completely ignore factors such as:

- Differing regularity systems for planning, building control
- Differing legal systems and contracts, procurement routes
- Differing expertise and methods of construction
- Quality of the finished product
- Differing attitudes towards health and safety issues
- Differing employment systems
- Specification levels
- Exchange rates
- Mandatory insurance requirements
- Environmental requirements
- Etc.

Table 1.1 Cost comparison for Central Business District Offices in US $

Location	Turn Out Cost US $	Labour Cost US $
Dublin	4,660.00	49.00
Paris	4,048.00	71.00
London	5,261.00	68.40
Melbourne	3,726.00	97.50
Hong Kong	4,085.00	23.00

Source: Turner and Townsend International Construction Market Survey 2022.

On the face of it, Table 1.1 appears to indicate that despite labour costs being higher in Melbourne than London turn-out costs are considerably cheaper and politicians could be forgiven, looking at the raw data, for coming to the conclusion that they are being ripped off.

In an attempt to make international comparisons more realistic, the RICS became involved with two initiatives, the International Property Measurement Standards (IPMS) and the International Cost Measurement Standards (ICMS).

International Property Measurement Standards (IPMS) Various Editions (2014–2019)

International Property Measurement Standards (IPMS) Various Editions (2014–2019) was drafted by an international coalition of surveyor's professional bodies in record breaking time. For international clients, building in several countries worldwide, it has always been problematic when try to compare prices and determine value for money for construction work. There were concerns about accuracy of current approaches when trying to measure and compare costs across various countries. The Building Cost Information Service (BCIS) found a lack of common classifications and a complete absence of standards in certain countries resulting in,

- Buildings are measured differently everywhere in the world
- Property measurements varying by as much as 24% (Jones Lang LaSalle) depending based on the approach to measurement adopted.

The drivers therefore of the IPMS and the ICMS are to:

- Ensure consistency of approach
- Capture good practice
- Reduce distortion when comparing costs in different markets
- Ensure transparency of data.

It is hoped that the IPMS will replace *The Code of Measurement Practice* (*6th Edition*) in due course.

The traditional approaches to comparing cross-border cost are as follows:

Convert to a single currency such as US $

This is the most common means of comparison, useful for a multinational organisation paying for projects in its home currency.

Advantages

- Easy to understand and visualise
- Gives the cost of typical building in each country.

Disadvantages

- A change in the exchange rate makes a significant difference: if a particular currency is strong compared to the base currency, the cost of construction looks expensive
- Is not a reliable indicator of relative costs and efficiency of construction between countries
- All countries/regions have individual nuances, for example, the rate of social security payments.

Location index

The location index offers a common base to compare costs of construction across different markets. London equals 100 as the initial reference point, with the deviation from 100 driven by average cost in US $ of several different building types of construction.

Each building category, at the time of writing, there are four, has it own unique set of measurement rules:

- IPMS for offices 2014
- IPMS, for residential buildings 2016
- IPMS for industrial buildings 2018
- IPMS for retail 2019.

Additionally, each category of IPMS comprises three sections:

IPMS 1 – which equates closely to the former GEA (gross external area)
IPMS 2 – which equates closely to the former GIA (gross internal area)
IPMS 3 – which equates closely to the former NIA (net internal area), sometimes also referred to as net lettable area, net usable area, carpet area, etc.

For the quantity surveyor, the most used set of rules will be IPMS2, typically used in the preparation of estimates and cost plans. The reaction of the profession to the new set of rules has been very similar to the introduction of the New Rules of Measurement Suite, with accusations of lack of consultation as well as questioning the relevance of the rules to everyday practice. Copies of the IPMS can be downloaded at https://ipmsc.org/standards/

International Cost Measurement Standards (ICMS) 3rd Edition 2021

Like IPMS, the ICMS is the product of a coalition of professional bodies across the world. The standards started life as the International Construction Measurement Standards in 2017 and was rebadged in 2021 as International Cost Measurement Standards. In response to industry feedback following the publication of the first edition of ICMS, the ICMS Coalition updated the standard in 2019 to incorporate life cycle costs. The second edition enabled practitioners to classify costs across the whole project life cycle, eliminating inconsistencies and discrepancies when accounting, comparing and predicting project finances. In 2021, recognising the importance of reducing carbon emissions in construction, the ICMS Coalition developed the third edition. This provides a common reporting framework allowing the inter-relationship between construction costs and carbon emission to be explored. The third edition provides the opportunity to make decisions about design, construction, operation and measurement of the built environment that optimise environmental sustainability.

The aim of ICMS is to provide global consistency in classifying, defining, analysing and reporting construction costs at a project, regional and national level. ICMS allows:

- Construction costs to be consistently and transparently benchmarked
- The causes of differences in costs between projects to be identified enabling properly informed decisions on the design and location of construction projects to be made.

It is important to note that ICMS will not replace existing measurement guidelines, such as the New Rules of Measurement (NRM) or Civil Engineering Standard Method of Measurement (CESMM), or bespoke client cost breakdown structures. Instead, they are a new way of presenting and reporting infrastructure costs to clients, stakeholders and investors at all project stages A copy of the ICMS3 can be downloaded at

https://www.rics.org/uk/upholding-professional-standards/sector-standards/construction/icms3/

Compared with many European countries, UK construction produces high output costs to customers from low input costs of professional, trade labour and materials. This fact is at the root of the Egan critique, pointing out that the UK has a wasteful system which would cost even more if UK labour rates were equal to those found in Europe. The waste in the system, ten years plus from Latham, is still estimated to be around 30%. Looking at French design and construction it is possible to see several of the Egan goals in place, but in ways specific to France. Whilst the design process begins without contractor involvement, they become involved sooner than in the UK and take responsibility for much of the detailed design and specification. They are more likely to buy standard components and systems from regular suppliers with well-developed supply chains, rather than on a project-by-project basis. Constructional simplicity follows from the French approach with French architects having little control of details and not appearing to worry too much about doors and window details for example.

Keeping the focus on Europe, for many observers, the question of single point or project liability – the norm in many countries, such as Belgium and France – is pivotal in the search for adding value to the UK construction product and is at the heart of the other construction industries' abilities to re-engineer designs. Single point project liability insurance, as recommended by Latham is insurance that protects all the parties involved in both the design and the construction process against failures in both design and construction of the works for the duration of the policy. The present system, where some team members are insured and some not, results in a tendency to design defensively, caveat all statements and advice with exclusions of liability, and seek help from no other members of the team – not a recipe for teamwork. In the case of a construction management contract, the present approach to latent defect liability can result in the issue of 20–30 collateral warranties, which facilitates the creation of a contractual relationship where one would otherwise not exist in order that the wronged party is then able to sue under contract rather than rely on the tort of negligence. Therefore, in order to give contractors, the power truly to innovate and to use techniques like value engineering (see Chapter 6), there has to be a fundamental change in the approach to liability. Contract forms could be amended to allow the contractor to modify the technical design prior to construction, with the consulting architects and engineers waiving their rights to interfere.

Opponents of the proposal to introduce single-point liability cite additional costs as a negative factor. However, indicative costs given by Royal & SunAlliance seem to prove that these are minimal – for example, traditional structural and weatherproofing: 0.65%–1.00% of contract value total cover, including structural, weatherproofing, non-structural and mechanical and electrical; 1%–2% of contract value to cover latent defects for periods of up to 12 years, to tie in with the limitations provision of contracts under Seal.

As in the French system, technical auditors can be appointed to minimise risk and, some may argue, add value through an independent overview of the project.

Coping with change

As resilient as quantity surveyors has proved to be over the years there is little doubt that many practices, both large and small, have perished because they failed to adapt to change, as discussed earlier in this chapter. Very often new innovation involves change, either in terms of the organisation or to personnel within the organisation and the quantity surveyor should be aware and respond to this. Change management refers to the process, tools and techniques to manage the people-side of change to achieve the required business outcome. On occasions a separate change manager may be appointed.

The change management cycle

Change management incorporates the organizational tools that can be utilized to help organisations and individuals make successful personal transitions resulting in the adoption and realization of change. Steps in the change management process are said to be (see Figure 1.7):

- Planning for change
- Managing change
- Reinforcing change.

In the widest sense, change management is a structural approach for moving organisations from their current state to a future state with anticipated business and organisational benefits. It helps organisations to adapt and align to new and emerging market forces and conditions. Delivery and handover of a successful project may well involve organisational change in order to get the maximum benefit from a project a well-managed handover is essential and organisations should be able to manage the process successfully.

The steps for an effective change management process are:

- Formulating the change by identifying and clarifying the need for change and establishing the scope of change
- Planning the change by defining the change approach and planning stakeholder engagement as well as transition and integration

Figure 1.7 Change management cycle.

- Implementing the change by preparing the organisation for change, mobilizing the stakeholders and delivering project outputs
- Managing the change transition by transitioning the outputs into business operations, measuring the adoption rate and the change outcomes and benefits and adjusting the plan to address discrepancies
- Sustaining the change on an ongoing basis through communication, consultation and representation of the stake holders, conducting sense making activities and measuring benefits.

A quantity surveyor can influence the culture to change by:

- Assessing stakeholder change resistance and / or support for change
- Ensuring clarity of vision and values among stakeholders
- Creating an understanding among the various stakeholder groups about their individual and interdependent roles in attaining the goals of the change initiative
- Building strong alignment between stakeholder attitudes and strategic goals and objectives.

Change management models

Lewin's change management model

This change management model was created in the 1950s by psychologist Kurt Lewin. Lewin noted that most people tend to prefer and operate within certain zones of safety. He recognized three stages of change:

- **Unfreeze** – most people make a conscious effort to resist change. In order to overcome this tendency, a period of thawing or unfreezing must be initiated through motivation
- **Transition** – once change is initiated, the company moves into a transition period, which may last for some time. Adequate leadership and reassurance is necessary for the process to be successful
- **Refreeze** – after change has been accepted and successfully implemented, the company becomes stable again, and staff refreezes as they operate under the new guidelines. While this change management model remains widely used today, it is takes time to implement. Since it is easy to use, most companies tend to prefer this model to enact major changes.

McKinsey 7-S model

The McKinsey 7-S model offers a holistic approach to organization. This model, created by Robert Waterman, Tom Peters, Richard Pascale, and

Anthony Athos during a meeting in 1978, has 7 factors that operate as collective agent of change:

- Shared values
- Strategy
- Structure
- Systems
- Style
- Staff
- Skills.

The McKinsey 7-S Model offers four primary benefits:

- An effective method to diagnose and understand an organization
- Provides guidance in organizational change
- Combines rational and emotional components
- All parts are integral and must be addressed in a unified manner.

The disadvantages of the McKinsey 7-S Model are:

- When one-part changes, all parts change, because all factors are interrelated
- Differences are ignored
- The model is complex and companies using this model have been known to have a higher incidence of failure.

Kotter's 8 step change model

Created by Harvard University Professor John Kotter, the model causes change to become a campaign. Employees buy into the change after leaders convince them of the urgent need for change to occur. There are 8 steps are involved in this model:

- Increase the urgency for change
- Build a team dedicated to change
- Create the vision for change
- Communicate the need for change
- Empower staff with the ability to change
- Create short term goals
- Stay persistent
- Make the change permanent.

Significant advantages to the model are:

- The process is an easy step-by-step model
- The focus is on preparing and accepting change, not the actual change
- Transition is easier with this model.

However, there are some disadvantages offered by this model:

- Steps can't be skipped
- The process takes a great deal of time
- It doesn't matter if the proposed changed is a change in the process of project planning or general operations.

Adjusting to change is difficult for an organization and its employees and using almost any model is helpful to change managers, as it offers leaders a guideline to follow, along with the ability to determine expected results.

Organisational development

Organisational development is technique to formalise approaches of organisations that are subject to continuous and rapid change. Ways of implementing organisational development include:

- Employing external consultants to advise on change
- Establishing an internal department to instigate organisational change
- Integrating the change process within the mainstream activities of the organisation.

There are a variety of opinions as to which approach is best as each has its strengths and weaknesses.

Business process re-engineering

Business process re-engineering on face value sounds very similar to organisational development and in practice the two approaches can be difficult to separate.

The idea of re-engineering was first propounded in an article in *Harvard Business Review* in July–August 1990 by Michael Hammer, then a professor of computer science at MIT. The method was popularly referred to as business process re-engineering (BPR) and was based on an examination of the way information technology was affecting business processes. BPR promised a novel approach to corporate change and was described by its inventors as a *fundamental rethinking and radical redesign of business processes to achieve dramatic improvements in critical measures of performance such as cost, quality, service and speed.*

The technique involves analysing a company's central processes and re-assembling them in a more efficient fashion and in a way that overrides long-established and frequently irrelevant functional distinctions. A similar approach is adopted by value engineering (see Chapter 6). Throughout this book, there is frequent reference to the traditional silo mentality of the construction industry. Silos that are often protective of information, for instance, and of their own position in the scheme of things. Breaking up and redistributing the silos into their different processes and then re-assembling them in a less vertical fashion exposed excess fat and forces organisations to look at new ways to streamline themselves.

One of the faults of the idea, which the creators themselves acknowledged, was that re-engineering became something that managers were only too happy to impose on others but not on themselves. Mammer's follow-up book was pointedly called 'Re-engineering Management'. 'If their jobs and styles are left largely intact, managers will eventually undermine the very structure of their rebuilt enterprises', he wrote with considerable foresight in 1994. BPR has been implemented with considerable success by some high-profile organisations; however, it has been suggested that construction due to its fragmented nature is a barrier to inter-organisational change.

Less adversarial contracts

Despite its many critics, for many years, the default contract recommended by quantity surveyors was the Joint Contracts Tribunal contract, known as the JCT. The main reason for this seems to be that everyone concerned in the construction process is familiar with the JCT, in all its forms, and more or less knows what the outcome will be in the event of a contractual dispute between the parties to the contract. However, the JCT was often blamed for much of the confrontation that has historically been so much a part of the everyday life in the construction industry and Latham in his 1994 report recommended the use of the NEC. The NEC was first published in 1993 with a second edition (NEC2), when it was renamed the Engineering and Construction Contract (ECC) and a third edition followed (NEC3) in July 2005 with a fourth edition in 2017. According to the RIBA/NBS Contracts and Law Report 2022, NEC contracts were used in 31% of contracts surveyed. The NEC is now mandated for many central government agencies, including the NHS ProCure22 and the Government Procurement Service recommends that public sector procurers use the NEC on their construction projects. NEC was used for Crossrail and Transport for London. One of the main differences between NEC and more traditional forms of contract is that the NEC has deliberately been drafted in non-legal language in the present tense which may be fine for the parties to the contract but can cause concern to legal advisors who must interpret its effect. Another innovation is the inclusion of a risk register which although makes possible the early identification of risks has led to concerns that it may be skewed in the contractor's favour, obliging the project manager to cooperate to the contractor's advantage. To date it appears as though the NEC is a step in the right direction for construction. There are few disputes involving the NEC that have reached the courts and there is no substantive NEC case law, but time will tell whether that continues to be the same when it becomes more widely adopted.

Building Information Modelling (BIM)

Construction Industry Institute research has identified that the average cost of re-work can be between 3% and 12% of project cost, no wonder then that BIM has been seen as an opportunity to get things right first time and save an awful lot of money and time. The Government Construction Strategy 2011, referred to previously, made several demands on the construction industry, perhaps the highest profile of which was that Level 3 BIM was to be mandatory on all public sector contracts by March 2016, irrespective of value. Wales and Northern Ireland followed suite with Scotland waiting until 2017 when BIM was required for all EU public procurement projects. The NBS Digital Construction Report 2021 – see Figure 1.8 indicates that in a survey of nearly 1,000 construction professional 71% of those surveyed have adopted BIM. That on face value, this is impressive, however, what does 'adoption' mean? For just under a third of respondents, it meant working with 3D models although of course, BIM is far more than that, it is about information management and working to a set of standards.

BIM is primarily an IT-based system that allows for greater collaboration between members of the demand and supply side (including quantity surveyors) of the industry at all stages in pre- and post-contract as well as operation. BIM is not new, it has been used in other market sectors for decades and also in other forms in the construction industry. Put simply, BIM is a collaboration platform that uses proprietary software to create a 3D model which can be accessed by all members of the construction team. BIM therefore

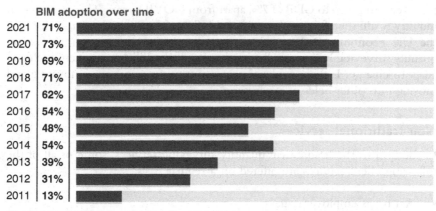

BIM adoption over time

Year	Adoption
2021	71%
2020	73%
2019	69%
2018	71%
2017	62%
2016	54%
2015	48%
2014	54%
2013	39%
2012	31%
2011	13%

Source: NBS Digital Construction Report 2021

Figure 1.8 NBA Digital Construction Report 2021.

Source: NBA Digital Construction Report 2021.

is capable of producing a model of a building prior to construction starting on site with the advantages that:

- Clashes between different elements can be detected and eliminated
- Alternatives can be investigated
- Time can be saved once work starts on site.

In addition, the following information can be loaded onto the model:

- Operation and maintenance data
- Life cycle cost data
- Health and safety details.

One thing is certain; the way in which BIM has been sold to the construction industry has been a disaster as in essence BIM is not all about software and IT systems – the real challenge for BIM is to encourage greater transparency and collaboration in the construction process. BIM will be discussed more fully in Chapter 4. In 2015, the RICS launched a BIM certified manager scheme for applicants from a wide range of professions, including those with 12 months of practical BIM experience either in cost estimating or construction.

There can be no doubt that the pressure for change within the UK construction industry and its professions, including quantity surveying, is unstoppable, and that the volume of initiatives in both the public and private sectors to try to engineer change grows daily. The last decade of the twentieth century saw a realignment of the UK's economic base. Traditional manufacturing industries declined while services industries prospered, but throughout this period, the construction industry has remained relatively static, with a turnover compared to GDP of 7%, apart from COVID years. The construction industry is still therefore a substantial and influential sector and a major force in the UK economy. Perhaps more than any other construction profession, quantity surveying has repeatedly demonstrated the ability to reinvent itself and adapt to change. Is there evidence that quantity surveyors are innovating and moving into other fields of expertise?

Non-traditional services

In addition to the 'traditional' quantity surveying services, some or all of the following client services are offered on a regular basis:

- Acting as employer's agent
- Acting as a project monitor
- Acting as a development manager
- Acting as a taxation advisor
- Working within infrastructure sector including the energy sector (oil and gas).

Employer's agent

The most common situation where an employer's agent is used is under the JCT (16) Design and Build contract. Design and Build and its variants has over the past 20 years overtaken traditional lump sums contracts based on JCT (16) as the popular form of contract amongst clients, with more that 40% of contracts being let on this basis in 2016. The employer's agent, as provided in JCT (16) D&B Article 3: Employer's Agent of the contract, is employed to administer the conditions of contract, but does not perform the same function as the architect, contract administrator or project manager.

The certifier

The role of the certifier is different to and separate from role of EA. The employer's agent has very little discretion in carrying out their duties. However, once the role extends to include issuing certificates or approvals and requires the exercise of discretion and professional expertise then the position becomes more complicated. As with general project management appointments, there is no commonly accepted standard role and service for the role of employer's agent. However, assuming a broad role both pre- and post-contract, the following could form the basis of an agreed role.

Project monitoring

Project monitoring is distinct from both project management and construction monitoring and is defined in the RICS Project monitoring guidance note (2007) as: Protecting the client's interests by identifying and advising on the risks associated with acquiring and interest in a development that is not under the client's direct control.

Development manager

The RICS Guidance Note on Development Management defines the role as

> *The management of the development process from the emergence of the initial development concept to the commencement of the tendering process for the construction of the works.*

The role of the development manager therefore includes giving advice on:

• Development appraisals
• Planning application process
• Development finance
• Procurement.

There are several definitions of the term development manager and there follows a comparison of the development management process as defined by:

The RICS Guidance Note
The CIOB's Code of Practice for Project Management for Construction and Development
Construction Industry Council (CIC) Scope of Services (major projects).

Infrastructure

Infrastructure covers a very wide area, including:

- Civil engineering
- Oil and gas
- Petrochemical
- Railways
- etc.

Oil and gas

Quantity surveyors, or cost engineers as they are often referred to in the oil and gas sector, are usually engaged at the preliminary stages in preparing cost estimates for the construction of oil rigs, refineries, laying of pipelines or a shutdown for maintenance. As is the case for the building sector, cost engineers must have a working knowledge of drawings, terminologies and the specific skills required for mechanical and pipe work. Often oil and gas facilities are linked to one another, and that any shutdown period needs to be planned carefully. Timing is crucial, whether the project is new-build, upgrade or maintenance. Services provided by cost engineers include the following:

- Cost estimates
- Tender documentation
- Pre- and post-contract administration
- Monthly valuations
- Financial reports
- Evaluating contractual claims
- Agreeing final accounts
- Project management
- Facilities management
- Technical audits.

For the quantity surveyor working as a cost engineer seeking to gain corporate membership of the RICS, the APC Built Infrastructure requires the candidate to select from a range of level 3 competencies that include:

- Energy: (utilities, renewable sources and nuclear)
- Mining and resources
- Oil and gas
- Petrochemicals
- Transport: including road, rail, aviation and ports.

Bibliography

Accelerating Change (2002). *Strategic Forum for Construction.*

Achieving Excellence (1999). *Modernising Construction,* Office of Government Commerce.

Banwell, Sir H. (1964). *Report of the Committee on the Placing and Management of Contracts for Building and Civil Engineering Work,* HMSO.

Burnside, K. and Westcott, A. (1999). *Market Trends and Developments in QS Services,* RICS Research Foundation.

Cabinet Office (2011). *Government Construction Strategy 2011.*

Callcutt J. (2007). *Review of House building Delivery,* The Home Builders Federation.

Carr, B. and Winch, G. (1998). *Construction Benchmarking: An International Perspective,* University College London, London.

Cook, C. (1999). QS's in revolt, *Building,* 29 Oct, p. 24.

CRINE (1994). *Cost Reduction Initiative for a New Era,* United Kingdom Offshore Operators Association.

Department of the Environment, Transport and the Regions (1998). *Rethinking Construction,* HMSO.

Emmerson, Sir H. (1962). *Survey of Problems before the Construction Industries,* HMSO.

Goodall, J. (2000). *Is the British Construction Industry Still Suffering from the Napoleonic Wars?* Address to National Construction Creativity Club, London, 7 July 2000.

HM Government (2013). *Construction 2025,* HM Government.

Hoxley, M. (1998). *Value for Money? The Impact of Competitive Fee Tendering on the Construction Professional Service Quality,* RICS Research.

Infrastructure and Projects Authority (2016). *Government's Construction Strategy 2016–20.*

Jones, Chow and Gibb (2020). *COVID-19 and Construction: Early Lessons for a New Normal?* Loughborough University.

Latham, Sir M. (1994). *Constructing the Team,* HMSO.

Lema, N. M. and Price, A. D. F. (1995). Benchmarking: Performance improvement toward competitive advantage, *Journal of Management in Engineering,* 11(1), pp. 28–37.

Levene, Sir P. (1995). *Construction Procurement by Government. An Efficiency Scrutiny,* HMSO.

McKinsey and Company (2020). *The Next Normal in Construction.*

National Audit Office (2001). *Modernising Construction,* HMSO.

NBS (2021). *Digital Construction Report 2021,* NBS Enterprises.

Office of National Statistics (2020). *Construction Statistics,* ONS, Great Britain.

Oxford Economics and Carpenter G (2022). *The Future of Construction.* Marsh Ltd.

RIBA/NBS (2022). *Construction Contracts and Law Report,* RIBA Publishing.

RICS (2013). *Black Book Cost Analysis and Benchmarking Guidance Note,* Royal Institution of Chartered Surveyors.

RICS (2015). *RICS Futures, Our Changing World: Let's Be Ready*, Royal Institution of Chartered Surveyors.

RICS (2019). *Future of the Profession*, Royal Institution of Chartered Surveyors.

RICS (2020). *RICS Futures 2020*, Royal Institution of Chartered Surveyors.

Royal Institution of Chartered Surveyors (1971). *The Future Role of the Quantity Surveyor*, RICS.

Royal Institution of Chartered Surveyors (1998). *The Challenge of Change: QS Think Tank 1998; Questioning the Future of the Profession*.

Simon, Sir E. (1944). *The Placing and Management of Building Contracts*, HMSO.

The Cabinet Office (2018). *Best Practice in Benchmarking*, HMSO.

Thompson, M. L. (1968). *Chartered Surveyors: The Growth of a Profession*, Routledge & Kegan Paul.

Turner and Townsend (2022). *International Construction Market Survey 2022*.

The International Cost Measurement Coalition (2021). *International Cost Measurement Standards 3rd Edition*, ICMC.

The International Property Measurement Standards Coalition (2014–18). *International Property Measurement Standards*, IPMS.

2 The challenges ahead

The coming years are going to require a range of new skills from quantity surveyors to meet the challenges of a rapidly changing industry and client demands. Fortunately, the quantity surveyor has proved, over the past 150 years, to be extremely resilient and receptive to new challenges as discussed in Chapter 1. Following the RICS Futures report, Our Changing World, the RICS conducted a series of conversations through rounds, seminars, consultation and outreach to understand what members thought were the challenges facing the profession. This exercise formed the basis of the Futures Report (2020) with 66% of surveyors consulted thinking that changing business models would impact on their future working practices.

The RICS Futures Report 2020 highlights three challenges facing the profession:

1 The increased importance of data and technology, discussed in Chapter 4
2 Sustainability
3 Providing the new talent and skills required to deliver the above.

Sustainability

Various attempts have been made to define the term 'sustainable or green construction'. In reality, it would appear to mean different things to different people in different part of the world depending on local circumstances. Consequently, there may never be a consensus view on its exact meaning; however, one way of looking at sustainability is '*The ways in which built assets are procured and erected, used and operated, maintained and repaired, modernized and rehabilitated and reused or demolished and recycled constitutes the complete life cycle of sustainable construction activities*'. In 2005, the RICS announced that it was establishing a new commission with a mission to; '*Ensure that sustainability becomes and remains a priority issue throughout the profession and RICS*'. The RICS Sustainability Report 2021 states the progress is mixed but tends towards the positive but not at the required rate. In general, a sustainable building reduces the impact on the environmental and social systems that surround it as compared to conventional buildings. Green buildings use less water and

DOI: 10.1201/9781003293453-2

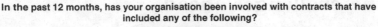

In the past 12 months, has your organisation been involved with contracts that have included any of the following?

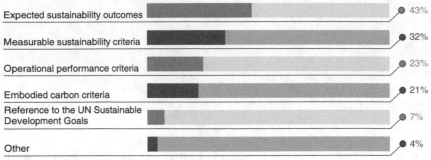

Figure 2.1 Sustainability as a part of project outcomes.

Source: RIBA Construction Contracts and Law Report 2022, https://www.architecture.com/knowledge-and-resources/knowledge-landing-page/riba-construction-contracts-and-law-report-2022.

energy, as well as fewer raw materials and other finite resources. The RIBA Construction Contracts and Law Report 2022 investigated the extent that sustainability is a part of project outcomes (see Figure 2.1). It is good to see that in 43% of responses, sustainable outcomes were expected to be part of project outcomes.

Value from green development

There now seems to be awareness that energy-efficient buildings command higher rental prices, have lower vacancy rates and higher market values, relative to otherwise comparable conventional office buildings. This, along with revenue savings of reduced energy and water as well as expert opinion that suggests higher levels of productivity can be achieved, begins to create a compelling argument for building and refurbishing to higher green standards. However, despite this, energy efficient buildings are still in the minority of total building stock. Emissions from buildings alone in 2020 accounted for 38% of total global energy-related CO_2 emissions and that is without counting emissions from the manufacture of materials. In addition, an estimated 25 million tonnes of construction waste end up in landfill without any form of recovery or reuse. Understandably, successive UK Governments have set out to reduce construction waste to landfill, for economic and environmental reasons. Construction 2025 sets out ambitious targets in four key measurable areas which included a 50% reduction in greenhouse gas emissions in the built environment. The strategy is part of the Government's wide-ranging industrial policy identifying key growth sectors in the economy, which has seen it launch similar visions for the automotive and oil and gas sectors. There is little wonder that the construction industry and its associated materials and manufacturing sectors has been singled out for action in the green debate when the statistics are laid out.

Comparatively few green buildings have been completed and of those a high percentage have been in the public sector. However, construction clients increasingly are realising the marketing potential of green issues for example:

- Significantly reduced whole life costs, as discussed in Chapter 3, and ensure more rapid pay back compared to conventional buildings from lower operation and maintenance costs, thereby generating a higher return on investment
- Secure tenants more quickly
- Command higher rents or prices
- Enjoy lower tenant turn over
- Attract grants, subsidies, tax breaks and other inducements
- Improve business productivity for occupants. 85% of a building's real costs are related to staff/productivity costs – user satisfaction is therefore key
- Image – branding and symbolising values
- 30% of newly built or renovated buildings suffer from sick building syndrome.

However, the barriers to green development are at present substantial and include the following:

- Lack of a clear project goals – i.e., targets
- Lack of experience
- Lack of commitment
- Complicated and expensive rating processes.

A raft of new legislation is now ensuring that the consideration of sustainability for new projects is not merely an option. In January 2006, the European Commission's Energy Performance Directive came into effect and shortly after that in April 2007 the Office of the Deputy Prime Minister launched the Code for Sustainable Homes. The Code for Sustainable Homes has now been withdrawn (aside from the management of legacy cases) and has been replaced by new National Technical Standards which comprise new additional optional Building Regulations regarding water and access as well as a new national space standard (this is in addition to the existing mandatory Building Regulations). These additional options can be required for planning permission. On 15 June 2022, an important update to the Building Regulations took place which included an amendment to Part L. All new homes must now produce 31% less carbon emissions than what was previously acceptable in the Part L regulations. The updated Building Regulations are one of the most far-reaching pieces of legislation ever to hit the construction industry and will force cuts in the carbon emissions from buildings. The targets set by this change are just an interim measure paving the way for more stringent changes in 2025 with the

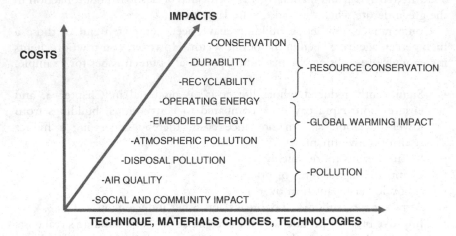

Figure 2.2 Techniques, materials choices and technologies.
Source: Atkins Faithfull and Gould.

introduction of the Future Homes Standard which will ban fossil fuel appliances in new homes in 'the shortest possible timeline'.

The process of getting the minimum life cycle cost and environmental impact is so complex, being a three-dimensional problem as indicated by Figure 2.2. Each design option will have associated impacts and costs, and trade-offs must be made between apparently unrelated entities, for example, what if the budget demands a choice between recycled bricks or passive ventilation.

The building passport

Building passports are being developed to provide a single point reference for all the information associated with a building throughout its life cycle. The building passport or property logbook provides a digital identity for buildings in a controlled and protected environment. Unlike BIM, as discussed in Chapters 1 and 4, the building passport provides a long-term identity and more information for the building for example, floor plans, legal matters and safety requirements, rather than performance.

Green and sustainable construction

The global green and sustainable building industry is forecast to grow at an annual rate of 23% between 2017 and 2025 as a result of increasing carbon regulatory requirements and greater societal demand for greener products according to research carried out by ARCADIS in 2022. Construction 2025

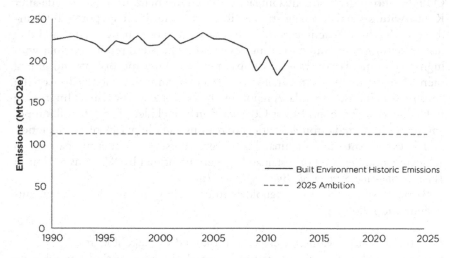

Figure 2.3 Actual built environment carbon emissions against 2025 reduction target.
Source: Green Construction Board.

set the industry a challenging target of a reduction in carbon emissions in the built environment of 50% by 2025. The ARCADIS research goes on to suggest that an increasing percentage of architects, quantity surveyors and contractors will see the majority or their work as green in the coming years. The strategy is part of the Government's wide-ranging industrial policy identifying key growth sectors in the economy, which has seen it launch similar visions for the automotive and oil and gas sectors.

During the preparation of previous editions of this book, sustainability and green issues were discussed by few in the construction industry during day-to-day business. Sustainability, a little like stress, appears to have crept up on the UK construction industry, over the past 20 years or so. And yet, concerns about climate change and the environment can be traced back several centuries, although it was not until late 1960s when organisations such as Greenpeace were formed that it rose to public attention. There followed a number of reports and protocols, the most notable being:

- The Brundtland Report (1987) also known as Our Common Future, which linked sustainability and development. It established triple bottom line sustainability, environment, social and economic forces
- The Rio Declaration (1992) which established the concept of *the polluter pays*
- The Kyoto Protocol (1997) agreed under United Nations Framework Convention on Climate Change
- The Paris Climate Agreement (2015).

Opinion about environmental impact gathered momentum in the decade after Kyoto with several severe warnings, including the Stern Report (2005) indicating that human activity was the primary cause of climate change and that urgent action was required to change behavior. The Stern report together with high-profile media coverage of campaigners blocking motorways and gluing themselves to underground trains has placed sustainability at the top of the construction industry agenda. A report by the National Audit Office; Improving Public Services Through Better Construction, concluded that £2.6 billion per annum is still wasted through a variety of reasons including; lack of consideration of life cycle costs and sustainability or green issues. Numerous barriers to achieving a higher level of sustainability were identified by Williams & Dair in *What Is Stopping Sustainable Building in England?*

Barriers experienced by stakeholders in delivering sustainable developments include the following

*   Sustainability measures were not considered by stakeholders
*   Sustainability measures were not required by client (includes purchasers, tenants and end users)
*   Stakeholder has no power to enforce or require sustainability measures
*   One sustainability measure is traded for another
*   Sustainability measures were restricted or not allowed by regulators
*   Sustainability costs too much and investor will not fund it
*   Stake holders lack information
*   Inadequate, untested, or unreliable sustainable options are proposed
*   Sustainable options are not available.

One of the major obstacles to the introduction of more sustainable design and construction solutions is the perception that to do so will involve additional costs – typically 10% on capital costs. To obtain a high BREEAM ratings, there is the need to incur significant up-front investment (See Chapter 3). Nevertheless, the general uncertainty over the cost impact to an entire development's profitability could deter cost adverse funders from backing a green project.

Ramboll's sustainable buildings market study 2021

Ramboll's biennial Sustainable Buildings Market Study reveals insights from architects, developers, contractors and investors across the Nordics, Germany and the UK.

*   90% of respondents believe that sustainability is an important part of a business strategy
*   However, barriers to adopting a more sustainable approach include; lack of financial incentives (57%), lack of technical solutions (36%) and lack of technical knowledge (35%)

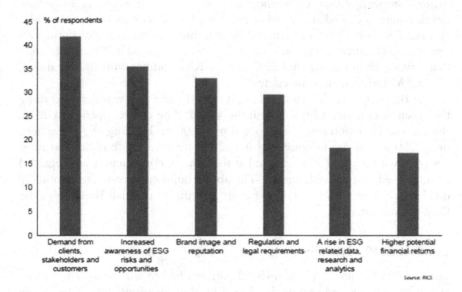

Figure 2.4 Factors driving green development.

- Surprisingly, 41% of respondents didn't know how much higher a rental yield can be commanded for a sustainable building and 37% didn't know how much higher sustainable property values will be
- Life cycle thinking, health and wellbeing and carbon neutrality were identified as the three most important trends, at 46%, 59% and 56% respectively
- 'Client, tenant or other stakeholder requirement', 'Enhancement of building performance' and 'Improved quality' were the most important drivers for using environmental certification schemes.

Whilst the findings show that sustainable buildings are becoming more mainstream, it also confirms a lack of clarity on the costs and benefits of sustainable buildings. When asked if sustainable buildings are a good investment, close to 50% of all respondents have little or no insight if sustainable buildings cost more to build, if they have reduced operational cost or if they trade at a premium. It is clear there is a lack of hard evidence to show whether sustainable buildings yield a positive return on investment. Addressing this knowledge gap will be vital for accelerating the uptake of sustainable buildings (Figure 2.4).

King's Cross redevelopment London

Over recent years, the area around King's Cross station in London has transformed into one of the most striking urban developments in the capital city. What was once an area of poverty and dilapidation is now home to a major European transport hub and significant new retail space. A 26-hectare development included offices 64%, residential 14%, culture and leisure 14% and retail 8% on a previously under used industrial site, much of the development was financed by green loans and it was announced in November 2021 that it was carbon neutral with 25% of the UK's buildings with an outstanding BREEAM rating within the estate.

The Bullitt Center in Washington DC (see Figure 2.5) was designed to be the greenest commercial building in the world. The centre, opened in 2016, reported construction costs for the six-story 4,800 m² building of £18 million, or £3750 per square m converted into 2022 figures. The building has been designed with a life of 250 years and at the time of construction was regarded as cutting-edge sustainable design. The above build costs compare favourably on the face of it with the £4000 per m² for the Leadenhall Building in the City of London.

London 2012 olympics

London 2012 required all commercial partners had to adhere to specific sustainability policies and criteria developed by the organising committee, these generally related to procurement, materials, waste management and ethical trading. The Olympics Delivery Authority (ODA) let around 7,000 direct contracts worth £6 billion which, together with its suppliers, formed supply chains of around 75,000 sub-contracts and accounted for more than two-thirds of the total spending on the games. The chief criteria were as follows:

- Tenders were considered against sustainability requirements, such as minimising embodied energy, responsible sourcing and designing out waste, separately from cost to form a 'balanced scorecard'
- The ODA had a specific procurement policy in which identified that delivery on time and to agreed costs as critical to ODA's success. Sustainability was addressed by requiring that whole life cost of product/service (e.g., including disposal) were taken into account and by including environmental measures in procurement
- Sustainability requirements were developed for each contract based on the ODA sustainability strategy using the targets in each area (water, waste, energy, emissions and materials) to inform key performance indicators for each contract. Before tendering for the building of new venues, the ODA works with designers to ensure that the design meets requirements – which then go to the contractor as a package to deliver against.

By considering sustainability throughout the decision-making process, the ODA found that in some cases, it was able to make savings. For example, the preferred concrete provider was both the cheapest and the best environmentally because they used (about 25%) recycled aggregate instead of more expensive conventional aggregate.

Figure 2.5 The Bullitt Center Washington, Copyright Joe Mabel 2014.

Birmingham 2022 commonwealth games

14 out of the 16 games venues for the Birmingham 2022 Commonwealth Games were existing facilities, and the games villages also made use of existing accommodation infrastructure. Alexander Stadium and Sandwell Aquatics Centre were the only venues to be refurbished and newly constructed, respectively, with both these projects aligning to existing needs and regeneration plans.

Recycled or reclaimed?

Recycled content is defined in ISO 14021:2016 which specifies requirements for self-declared environmental claims, including statements, symbols and graphics, regarding products. It further describes selected terms commonly used in environmental claims and gives qualifications for their use. This International Standard also describes a general evaluation and verification methodology for self-declared environmental claims and specific evaluation and verification methods for the selected claims.

Recycled content is the proportion, by mass, of recycled material in a product or packaging. Only pre-consumer and post-consumer materials shall be considered as recycled content, consistent with the following usage of the terms:

• Pre-consumer material: Material diverted from the waste stream during a manufacturing process. Excluded is reutilization of materials such as rework, regrind or scrap generated in a process and capable of being reclaimed within the same process that generated it
• Post-consumer material: Material generated by households or by commercial, industrial and institutional facilities in their role as end-users of the product, which can no longer be used for its intended purpose. This includes returns of material from the distribution chain.

The use of recycled materials is already a requirement for a number of construction clients as follows:

• The Greater London Assembly (GLA) has drawn up its Whole Life-Cycle Assessments Guidance, which will likely influence the embodied carbon targets set by other local authorities throughout the UK
• As part of its carbon neutrality programme, Beard Construction purchased a 180-acre woodland on the edge of Bristol in 2022 to offset some of its environmental impact with carbon sequestration
• The UK Green Building Council's (UKGBC) Whole Life Carbon Roadmap has created a common vision and agreed actions for achieving net zero carbon in the construction, operation and demolition of buildings and infrastructure

- The Olympic Delivery Authority adopted minimum standards of at least 20% (by value) of materials used in the permanent venues, to be from recycled content for London 2012
- Land Securities and British Land are both targeting 2030 for net zero portfolios and have introduced strategies to reduce emissions from existing and new buildings
- The DfE is now requiring contractors to calculate and report the embodied carbon content of new schools
- The Welsh Government has set targets for their schools, and the MoD is applying the net zero definition to its entire building stock
- Minimum recycled content standards have been adopted for regeneration by Southwest England and Yorkshire Forward Regional Development Agencies, Leeds Holbeck and Raploch Urban Regeneration Company.

In many cases, capital costs can be saved by using existing materials.

Recycled materials are any materials that have been taken from the waste stream and re-processed and re-manufactured to form a part of a new product. This can include, according to the Waste and Resources Action Programme (WRAP), the following examples of higher recycled content products available at no extra cost.

Table 2.1 Examples of higher recycled content products available at no extra cost

Component	Typical Product	Products with Higher Recycled Content
Dense blocks	Brand A – 0% recycled content (£5.50/m^2)	Brand B – 50%–80% recycled content (£5.50/m^2)
Concrete roof tiles	Brand C – 0% recycled content (£550/1000)	Brand D – 25% recycled content (£550/1000)
Glass/mineral wool insulation	Brand E – 10% recycled content (£3.50/m^2)	Brand F – 80% recycled content (£3.00 m^2)

It is thought that material types that offer higher levels of recycled content than the ones specified in Table 2.1 are as provided in Table 2.2.

Table 2.2 Material types that offer higher levels of recycled content

Bulk aggregates	Ready mix concrete
Asphalt	Drainage products
Pre-cast concrete products	Concrete tiles
Clay facing bricks	Lightweight blocks
Dense blocks	Plasterboard
Ceiling tiles	Chipboard
Insulation	Floor coverings

Source: Wrap.

Table 2.3 Examples of reclaimed and recycled materials

Reclaimed Materials	Recycled Materials
Re-used timber or floorboards	Panel products with chipped recycled timber
Bricks cleaned up and re-used	Bricks crushed for hardcore
Steel section shot blasted and re-assembled	Percentage of recycled steel

Reclaimed materials are any materials that have been used before either in buildings, temporary works or other uses and are re-used as construction materials without reprocessing. Reclaimed materials may be adapted and cut to size, cleaned up and refinished but they fundamentally are being re-used in their original form (Table 2.3).

Figure 2.6 illustrates some examples where using reclaimed in place of new (even in place of new steel with a 40% recycled content) has reduced the environmental impact by up to 96%.

Analysis of Figure 2.6 clearly demonstrates that in the case of first fix timber and steel, the impact on the environment when choosing to use reclaimed materials is dramatic.

Cost of reclaimed materials

Currently, much of the reclamation industry is set up to meet the needs of small to medium-size builders. For larger projects, there is no established supply chain that can reliably supply builders with large quantities of re-claimed materials. Therefore, to undertake signature projects requiring large quantities of reclaimed materials inevitably results in long lead-ins and perhaps the establishment of storage for reclaimed materials as they become available (Table 2.4).

Costs of alternative (renewable) materials

Renewable construction materials are made from plant-derived substances that can be produced repeatedly. In contrast, most other construction materials are derived from raw materials which we cannot replace; oil, minerals, metals etc. Renewable construction materials range from very 'natural' unprocessed materials such as straw bales used for walling, to more refined products such as floor coverings manufactured using renewable polymers. Renewable materials are beginning to be introduced into mainstream construction for example, hemp and lime walling material, and it is claimed may soon become com-monplace in UK buildings.

Renewable construction materials require less embodied energy to manu-facture, process and transport them to their point of use. Embodied energy includes energy used in obtaining the raw materials (e.g., sand and gravel or

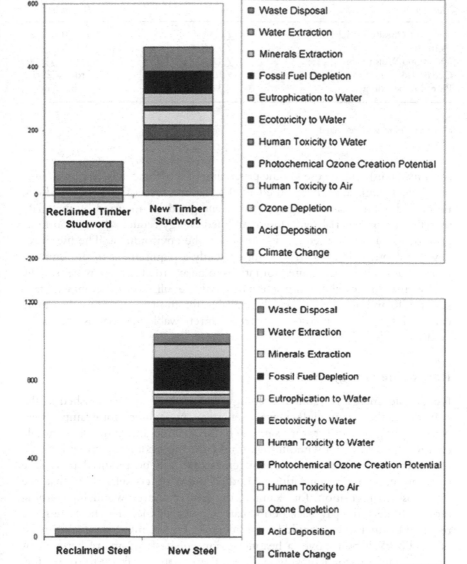

Figure 2.6 Life cycle analyses comparing new with reclaimed materials.
Source: BRE.

Table 2.4 Comparative cost of new and reclaimed materials excluding VAT, 2022 prices

Material	New	Reclaimed*
Wire cut Cheshire bricks	£0.80	£1.00
Flettons	£1.16	£0.75
500 × 300 Welsh roofing slate	£4.00–£6.00	£2.50–£3.00
Cedar cladding	£100.00/m^2	From £25/m^2
Pine floorboarding	£31.00/m^2	£35.00/m^2

Note
* Quantities may be restricted.

plant materials), energy used in the processing of those materials (e.g., grinding, blending, firing), and energy used in the manufacture of finished products. Embodied energy can be regarded as an energy 'debt' incurred by materials, therefore using materials with low embodied energy reduces the carbon footprint of a building and reduces its impact on the environment. The use of renewable construction materials is not currently a requirement in the Building Regulations and is not required for ratings to be awarded. This may be partially because many renewable construction materials are still at the development stage in the UK and are not yet widely available for most applications. For these reasons, information on costs associated with renewable materials is not widely available.

Renewable energy schemes

Renewable energy is defined as energy flows which are replenished at the same rate as they are used. Renewable energy may be direct, for example, solar water heating or indirect, for example, biomass, wind and hydro. Renewable energy schemes are not without their critics, the most common criticism being the sheer scale of say wind farm schemes that will be required to replace conventional power generation. Another point of contention is that the extra cost of installation, for example; the cost of a large woodchip burning biomass boilers, the pay-back period can be considerable, plus the extra space required for fuel storage could be considerable. For example, the fuel demand of 20,000 kWh boiler over a heating season equates to 2 m^3 of kerosene as against 25 m^3 of woodchips. In addition, there is the necessity to source fuel from within a 25-mile radius of the boiler to maintain a satisfactory carbon footprint. Against this the running cost of woodchip is substantially below the cost of conventional fuels, perhaps half in terms of cost per kWh. The price of an installation varies depending on the site specifics, but for the supply and installation of all the required biomass equipment in the plant room, the typical costs for a pellet/chip system are shown in Table 2.5.

Perhaps the most widely used applications of renewable energy sources are solar and photovoltaic panels. Used most often for domestic scale schemes, the

Table 2.5 Typical costs for a biomass pellet/chip
system

Rating	Approximate Cost £
50 kW	32,500.00
75 kW	37,500.00
100 kW	42,500.00
150 kW	50,000.00
200 kW	80,000.00
350 kW	125,000.00
500 kW	170,000.00
1 MW	250,000.00

main limiting factor is the area of panels required to produce energy. For example, in the case of solar panels for a 3-bedroom house, the cost will be around £5520–£6040 to install a 3–4 kW system to meet the higher electricity demands. The pay-back period for solar panels and photovoltaic panels is again considerable and could conceivably be longer that the design life of the project. Planning consideration may also be an issue.

Therefore, for many people, the most obvious signs of renewable energy schemes are wind turbines and wind farms. The cost of installation is considerable, but is capable of being off-set if the electricity produced is suitable for connection to the national grid, for example, £70,000 for an entry-level turbine callable of producing 15 kW-h. Other considerations are tax allowances and of course the suitability of the site for locating a wind turbine.

Ska ratings

It is estimated that 11% of UK construction spending is on fit-outs and that buildings may have 30–40 fit-outs during their lifecycle. Launched on 6 November 2009, the Ska rating is an environmental assessment method, benchmark and standard for non-domestic fit-outs, led and owned by RICS and compare the environmental performance of fit-out projects. Ska ratings are similar to BREEAM credits (see Chapter 3) but solely focuses on assessing a fit-out. In the case of a BREEAM Offices fit-out assessment, the tool filters out the land use and ecology credits and some of the credits relating to the build construction, therefore tailoring it for a fit-out assessment. However, some of the BREEAM Offices fit-out assessment credits do relate to the building which can be outside the control of the project, for example, whether the building has a pulsed water meter is assessed in BREEAM but the Ska rating only considers a water meter if the provision or modification of water services is within the scope of the fit-out. In addition, the BREEAM fit-out assessment has credits relating to the proximity of the building to public transport nodes (Ska rating does not assess this). The measurement practice that makes up the SKA assessment criteria covers eight different categories:

- Energy
- CO_2 emissions
- Waste
- Water
- Pollution
- Transport
- Materials
- Wellbeing.

Projects can achieve a Ska Bronze, Silver or Gold rating.

The Ska assessment process is broken into three stages:

- **Design/Planning.** At this stage, identify the measures and issues in scope. Once the measures in scope are identified the client has the opportunity to prioritise which measures, they want to achieve and make a decision against design, cost, programme and benefit, and add them into the scope of works for the project. This will also set the environmental performance standards for how the project is delivered, in terms of waste and energy in use etc. Then if the specification demonstrates that these measures are likely to be achieved, they will be reflected in an indicative rating
- **Delivery/Construction.** This involves the gathering of evidence from operating & maintenance manuals and other sources to prove that what has been specified has actually been delivered and that the performance and waste benchmarks have been achieved
- **Post-Occupancy assessment.** Finally, there is the option to review how well a fit out has performed in use against its original brief from a year after completion.

The RICS is currently operating an accreditation scheme for Ska assessment. For developers, the benefits of a Ska assessment are to make refit developments more attractive for tenants.

The construction industry and the circular economy

The circular economy can be summarised into three key principles:

1   Designing out waste and pollution
2   Keeping products and materials in use
3   Regenerating natural systems.

Circular construction seeks to eliminate waste production at all stages of the build process, from procurement and design through construction and into operation and then eventual end of life destinations. Circular construction also seeks to reduce the demand for virgin materials by keeping products and materials in use for as long as possible and using recovered materials.

Construction projects are very often cost and time prioritised, which can lead to working habits being wasteful and irrational for those tasked with producing the building. It is an undeniable fact that construction is very wasteful. According to several sources, the sector uses between 400 and 450 million tonnes of material every year, which results in 100–120 million tonnes of waste being produced and accounts, together with running costs, for 40% of all the UK's CO_2 emissions, compared with 2.50% for aviation. In addition, 30% of all road freight in the UK is building materials. To put this in context, this level of waste in construction contributes over a third of the UK's total yearly waste amount. Waste comes in a variety of forms from excavated materials to incorrectly order materials to packaging.

The traditional approach adopted in construction is the linear model, where ultimately raw materials are converted into waste. Typically, this involves existing buildings stripped to their shell and demolished, whilst a new re-placement structure on the existing footprint and fitted out with fresh materials. This model argues that any materials utilised in a building can only be used once before being disposed of as this produces revenue. It fails to recognise the fact that the supply of raw materials is finite as well as the excessive use of raw materials on the environment as illustrated in Figure 2.7.

A circular economy is said to one where production and consumption involves sharing, leasing, reusing, repairing, refurbishing and recycling existing materials and products as long as possible and appears to be a helpful solution to diminish the environmental impact of the industry.

The three principles required for establishing a circular economy are:

1   Eliminating waste and pollution
2   Circulating products and materials
3   The regeneration of nature.

A circular economy as opposed to the more conventional linear economy in which natural resources are transformed into products which are ultimately destined to become waste because of the way they have been designed and manufactured. It follows in a circular economy that building will have a longer life cycle than traditionally has been the case, this can be achieved by considering at an early stage, for example:

•   Deconstruction
•   Material recycling
•   Modular construction systems.

Take ➡ Make ➡ Use ➡ Dispose

Figure 2.7 Traditional linear model.

The main causes of the construction's environmental impact are found in the consumption of non-renewable resources and the generation of contaminant residues, both of which are increasing at an accelerating pace.

The circular economy can focus on:

- The operational (connected with particular parts of the production process)
- Tactical (connected with whole processes)
- Strategic (connected with the whole organization) levels.

Figure 2.8 illustrates an alternative approach to the traditional linear approach.

End-of-life buildings can be deconstructed, thereby creating new construction elements that can be used for creating new buildings and freeing up space for new development. Modular construction systems can be useful to create new buildings in the future and have the advantage of allowing easier deconstruction and reuse of the components afterwards (end-of-life buildings).

An interesting approach to reuse part or whole buildings is the concept of 'building in layers', first proposed by Frank Duffy in the 1970s and then

Figure 2.8 Circular economy and construction.

Source: Joslyn Institute.

20 years later developed by Stewart Brand. Each of the separate yet inter–linking layers, it is said, has different life spans as illustrated in Figure 2.9, therefore it becomes unnecesary to throw the baby out with the bath water when one of the layers needs replacement. The life span of the various layers varies between adoptors.

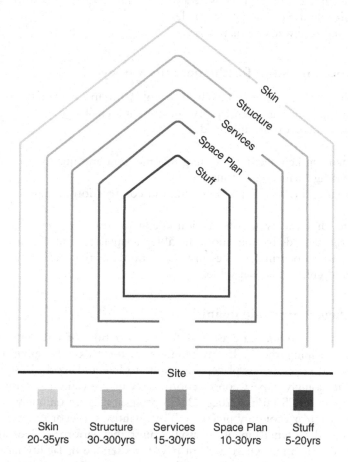

Skin	Structure	Services	Space Plan	Stuff
20-35yrs	30-300yrs	15-30yrs	10-30yrs	5-20yrs

Figure 2.9 Building in layers.

Source: Stewart Brand. How buildings learn. What happens after they are built.

Site: infinite, the fixed location of the building.

Structure: the building's skeleton including the substructure and load–bearing elements.

Skin: the façade and exterior.

Services: pipes, wires, energy and heating installations.

Space plan: the solid internal fit–out including walls and floors.

Stuff: the rest of the fitout including furniture, lighting and ICT.

What's in it for construction?

As with so many new initiatives the construction industry does not readily adopt new approaches to working however the perceived benefits of adopting a more sustainable approach can be seen as:

- Increased brand value/corporate image and reputation
- Cost savings through consideration of life cycle costs through the sustainable use of energy, water, less waste
- The use of fewer raw materials.

What does this mean for the quantity surveyor?

As sustainability is a comparatively new topic for most quantity surveyors, measuring and costing green factors can present a challenge.

Quantity surveyors are involved in:

- Influencing key decisions relating to time, cost and quality
- Managing finances and contracts
- Managing the design, procurement and construction process.

Therefore, they can assist construction clients to adopt new circular business models through delivering more flexible, adaptable, and deconstructable buildings and procuring services instead of products that offer new revenue streams and greater residual value.

RIBA stage 0 Strategic definition

Making an investment case is vital to the adoption of circular economy principles, helping the built environment sector tackle its environmental footprint and create better places for people to live and work. Business customers increasingly expect construction to act in a socially responsible way. According to RICS Futures 2020, RICS professionals are uniquely placed to introduce circular economy practices. The quantity surveyor can help clients to consider the ownership of an asset against the performance of an asset. Consider the situation where a client pays for a service or facility rather than taking ownership of a physical asset. The ownership of the built asset remains with the developer or producer meaning it is in the producer's interest to facilitate take-back and disassembly to maximise the value and utility of materials and components. If the project involves the reuse of an existing built asset, budget allocations can be agreed between savings relating to the reuse of products and materials and materials that may require maintenance, refurbishment or replacement during operation. In conjunction with facilities management, it may be prudent to allow a contingency for to cover this. To ensure a project has a focus on waste prevention from the outset and drive better performance, there is a need to establish a robust, transparent financial forecast in relations to the volume of waste predicted in the waste forecast. This initial forecast should be developed at the design stage and designing out waste activities should be undertaken. The quantity surveyor's role is to help to estimate the waste quantities using the bill of quantities documentation and refine this forecast by asking the different sub-contractors who supply materials and labour to site to give estimates of the waste that they will produce. Comparisons of actual waste produced vs. forecast waste should be collated to help inform future forecasts and to determine best practice levels which should be strived for. The true cost of wasted materials is on average ten times that of the waste disposal costs.

RIBA Stage 1 Preparation and brief

Above all, buildings must be affordable and constructed at an economic cost which people are prepared to pay. Sustainable development is vital but must be balanced against longer-term economic issues; this is the challenge faced by the quantity surveyor. Therefore, highlighting the economic benefits of alternative circular models on the project's lifecycle and operational expenditure costs is a crucial part of the quantity surveyor's role. Supporting the use of circular business models can also include incorporating alternative models of procurement which can reduce initial capital expenses and transfer these costs into the operations over a longer period through supply, maintain and replace contracts. When it comes to selecting the procurement route, make sure the criteria for reviewing selection take into account a contractor's track record of reuse of materials and experience in disassembly. The contract documentation should include a requirement for the use of re-used materials. There are two approaches:

1   The project reclamation target is built into all contracts allowing contractors the flexibility to meet the target as they see fit, for example,
    Reclaimed materials shall make up at least 5% of the total project materials by value as measured against the value of all construction materials used on the project.

2   Reclaimed products are specified for certain elements.

There are a number of models that can be used for procuring reclaimed materials that have worked successfully in the past, these are:

- Client buys and materials are sold to the contractor
- Contractor buys and all risks rest with the contractor
- Contractor and client buy; bulk easily sourced materials are purchased by the contractor with the client purchasing specialist materials
- Seek advice from reclamation specialists where appropriate, including at Concept and Technical Design stages and during construction

- Quantify potential reused and recycled content
- Engage with project insurers at Concept and Technical Design stages to work through concerns about use of specific reclaimed components
- If demolition contractor is being used specify in demolition contract which materials are to be reclaimed for reuse on site
- Also specify maximum wastage levels and establish inspection and audit trail for on-site materials.

RIBA Stages 2,3,4 Design

All sectors of the industry need to be committed to the concept of sustainable development if it is to be implemented successfully. A key element in this process is matching the procurement of materials to the objectives of stakeholders such as building occupiers who are concerned about the sustainability of the components that make up their property. The quantity surveyor is very influential at all stages of the property cycle throughout the property supply chain and supported by stakeholders, should be able to identify the cost of more durable alternatives, and the return on investment. Having this analysis, it is possible to increase the life of the product chosen in the design, not focusing this choice purely on its initial capital costs. The lifetime positive and negative impacts of a building on society and the environment from its construction, use, maintenance and repairs, decommissioning and disposal need to be recognised and accounted for. The liability for the performance of reused materials and products should be established and agreed. It is suggested that responsibility of the product or materials should rest with the client and responsibility for installation should rest with the contractor.

At the design stage, the quantity surveyor needs to be aware of the drivers for sustainability and the impact these have on capital and life cycle costs, as well as the technical requirements of sustainable buildings, so that these are developed into realistic costs and not arbitrary percentage additions. When the surveyor is required at this stage to liaise with the client and professional team to determine the client's initial requirements and to develop the client's brief, consideration should be given to the client's overall business objectives, particularly any corporate responsibility targets likely to affect the project. In advising the client on demolition and enabling works, the surveyor is advised to consider carrying out a pre-demolition audit to maximise material reclamation and reuse and minimise waste to landfill. The procurement of demolition and enabling works could include evaluation criteria that consider a company's sustainability credentials. Specialists should be required to contribute to meeting the client's objectives and the project targets in the key sustainability areas. Where the activities relate to CDM Regulations surveyors need to be aware of any Site Waste Management Plans. Where the client's objectives include achieving ratings/levels under BREEAM, LEED or the new National Technical Standards, surveyors would be expected to familiarise themselves with the specialists that need to be appointed both to carry out the assessment and to provide the necessary reports required by the schemes.

When advising on the cost of the project, sustainability implications of alternative design and construction options need to be understood. It is recommended that cost estimates include cost/m^2 information for indicative low and zero carbon and renewable energy schemes and material selection. Costing of issues not generally associated with building design is extremely important, for example, those actions identified in an Environmental Impact Assessment or the implications of a green travel plan, and the quantity surveyor would be expected to understand or be able to undertake life cycle assessment for the whole development not just the building. A site visit can identify issues likely to affect cost, time or method of application, including existing buildings on site, existing ecological features on site that may need protecting to achieve BREEAM credits, local road layouts that could create traffic congestion and noise, existing watercourses and the implications for storm water control and attenuation and areas of the site liable to flooding. Advising on the likely effect of market conditions can involve looking at the possible level of employment and skills in the area, and the levels of crime that might affect the site. The project costs at this stage can influence a financial appraisal and surveyors are advised to ensure they understand what is to be priced to provide a level of accuracy and avoid substantial cost increase later. In addition to considering the effects of site usage, shape of the building, alternative forms of design, procurement and construction, etc. the surveyor would be expected to be able to proactively advise on the sustainability implications of various low and zero carbon technologies, renewable energy installations and material selections. The surveyor would also be expected to be able to advise on the cost implications of other sustainability issues, including possible construction waste, levels of local employment and skills, traffic and transport.

When advising on tendering and contractual procurement options, it is recommended that consideration is given in pre-qualification documentation to evaluation of the bidder's response to sustainability issues, particularly those affecting the project. It is important to ensure that the client's and project's sustainability requirements that were incorporated into the project brief have been reflected in the tender documents and to ensure that the documentation also includes a responsible approach to sustainability in the contractor's operations, preliminaries and temporary work. Where bills of quantities are required, it is important that every effort is made to adequately measure all sustainability-related products and technologies, avoiding where possible provisional sums. It is important to ensure that the tender report identifies the sustainability issues/risks affecting the project and the bidder's response to them. An analysis of contractor's sustainability costs should be carried out to compare it with benchmarks and to report on this. It is important to ensure that variations with sustainability implications are valued and agreed.

It is commonly assumed that consideration of sustainable issues will rack up the costs of a building, but this may not necessarily be the case. One of the principal barriers to the wider adoption of more sustainable design and construction solutions is the perception that these incur additional unwanted costs. Location and site conditions have a major impact on the assessment and of course these factors may be outside of the design team's influence.

Key to achieving best whole life value:

- Understanding value
- Assessing value
- Putting a cost to value propositions
- Identifying the best value sustainable solution prior to commitment to invest
- Optimising value over the whole asset life cycle.

From a quantity surveyor's perspective, it must be possible to quantify and cost the impact of introducing sustainable practice into construction.

RIBA Stage 5 Manufacturing and construction

Increase product recycled content and reusability. In this model, opportunities are taken by tackling waste materials by either reusing materials before they become waste or create higher value uses through upcycling into new products or closing the loop by reincorporating the waste into the original product and therefore increasing the recycled content. These approaches can help to reduce primary material demand, avoid expensive, inefficient waste disposal and can often provide localised community benefits in many cases. Examples of this include:

- Implementing takeback schemes to return excess materials for resale as a typical mixed waste skip can contain up to 13% of unused materials which are perfectly reusable according to previous studies
- Working with waste management companies to secure supplies of materials which can be turned into new products – as demonstrated by Kenoteq Bricks made from construction and demolition waste although it must be said that the unit costs of alternatives to conventional materials are very uncompetitive. Another example is walnut husks being used for cleaning brick surfaces. Abrasive grains are produced from crushed, cleaned and selected walnut shells. They are classified as reusable abrasives
- Providing reusable packaging instead of single use such as reusable pallets instead of the wooden ones, reusable tarpaulins instead of shrink wrap and metal stillages for glazing deliveries instead of wooden ones. These measures will still ensure products arrive safe and intact but reduce packaging material costs for the manufacturer and disposal costs for the customer.

RIBA Stage 6 Handover

- Arrange a lessons-learnt session to establish which elements where reused
- The methodology for proposed deconstruction should be included in the operation and maintenance manual. If a BIM model has been produced, then this should be available
- If the building is to be let, then consideration should be given making it a contractual obligation that tenants inform the asset owner if changes are to be made to the building
- Soft landings should be introduced in order that the client/owner can maximise the efficient usage of the new project.

RIBA Stage 7 Use

- Develop a plan for the re-use of materials.

Talent and skills requirements

It should not be forgotten that measurement remains the primary skill of quantity surveyors. Paradoxically, almost a decade after the release of NRM2, many quantity surveyors and clients remained wedded to SMM7. In addition, matters have been further complicated with the publication of International Cost Measurement Standards and International Property Measurement Standards both of which seem to have been largely rejected by the rank-and-file RICS membership, see Chapter 1. The consultation following the RICS Futures Report concluded that the current recruitment to the profession was too narrow and more should be done to encourage non-traditional education routes including learning on the job apprenticeships as employers and school-leavers are questioning the value of a standalone degree course compared to structured learning and qualification while employed.

Accreditation for courses should be widened and courses that are accredited should be more flexible, reflecting future trends including artificial intelligence (AI), blockchain, big data and automation of many surveying tasks. The competencies for quantity surveyors should be constantly reviewed to reflect changing business models with crossovers between the pathways to produce more general multi-skilled broad-based professionals. It was also concluded that CPD will become even more important in the years to come. Finally, the House of Commons Environmental Audit Committee concluded in its publication 'Building to net zero: costing carbon in construction' (2022) argued that the 'chopping and changing of UK Government policy has inhibited skills development in housing design, construction and in the installation of new measures'. The committee heard repeatedly that skills gaps remain, inhibiting the industry from implementing low-carbon construction solutions.

Bibliography

ARCADIS (2022). *The Business Case for Intelligent Buildings*, Arcadis.

Arup (2016). *The Built Environment: From Linear to Circular*, Arup.

Arup/The Ellen MacArthur Foundation (2020). *From Principles to Practice: Realising the Value of Circular Economy in Real Estate.*

Brand, S. (1997). *How Buildings Learn. What Happens After They Are built*, W&N.

Brundtland, H. (1987). *The Brundtland Report, Our Common Future*, World Commission on Environment and Development.

CBI (2020). *Fine Margins: Delivering Financial Sustainability in UK Construction*, CBI.

Department of the Environment, Transport and the Regions (2000). *Demolition and New Building on Local Authority Estates*, Detr.

HM Government Construction 2025 (2013). *Industrial Strategy: Government and Industry in Partnership*, HMSO.

House of Commons Environmental Audit Committee (2022). *Building to Net Zero: Costing Carbon in Construction*, HMSO.

Joslyn Institute (2021). *The Future of Sustainable Design and Construction*, The Joslyn Institute.

Kelly, M. (2020). *Circular Economy 'Beginners' Checklists – Quantity Surveyors*, Southern Waste Region/Build360 Group.

Killip, G. (2020). A reform agenda for UK construction education and practice. *Buildings and Cities*, 1(1), pp. 525–537.

National Audit Office (2005). *Improving Public Services through Better Construction*, The Stationery Office.

Ramboll Buildings (2021). *Ramboll's Sustainable Buildings Market Study 2021.*

RIBA/NBS (2022). *Construction Contracts and Law Report 2022*, RIBA.

RICS (2020). *The Futures Report*, RICS.

Williams, K. and Dair, C. (2007). *What Is Stopping Sustainable Building in England? Barriers Experienced by Stakeholders in Delivering Sustainable Developments*, Wiley Online Library.

Wrap (2014). *Reclaimed Building Products Guide*, Waste & Resources Action Programme.

Zehra, K. (2021). *The RICS Sustainability Report 2021*, RICS.

3   Procurement

Almost 30 years since Latham and Egan called for a fresh approach to construction procurement with greater collaboration and transparency, it doesn't look like much has changed and in fact in the period since the fourth edition of *New Aspects of Quantity Surveying Practice* in 2018, the industry appears to have flat lined when it comes to attitudes on procurement and tendering. According to the RIBA Construction Contracts and Law Report 2022, traditional procurement, that's to say where client together with professional advisors sit on one side of the fence with contractors on the other, remains popular (see Figures 3.1 and 3.2). The approach is called traditional for a reason as the polarisation of roles means that collaboration and sharing of mutual objectives is as minimal as it was when Latham published 'Trust in the team'. When it comes to tendering, single stage competitive tendering with contracts being awarded to the first past the post is still out front (see Figure 3.2). To achieve better value, models of value must be defined which are broader than just bottom-line capital cost.

The RIBA Construction Contracts and Law Report 2022 contains a snapshot of current procurement practice. Procurement routes such as management contracting, construction management, measured term, cost plus and partnering/alliancing are still minority procurement routes being used most often by around 4% of respondents. The survey illustrates that traditional procurement was used most frequently in 56% of projects compared with 45% in 2018 (see Figure 3.1). Although Figure 3.1 appears to reinforce the march of design and build, there is anecdotal evidence that suggests that as the recession took hold in 2009 clients were reverting to single stage competitive tendering based on bills of quantities. One of the main reasons cited for this was the poor-quality control associated with design and build. Another interesting point about Figure 3.1 is the wide variation in some of the responses, in relation to construction management and partnering.

Public and private sector clients broadly agree what they want from the industry are:

- Project outcomes which meet their business case objectives
- Adequate data on which to make informed decisions e.g., early cost accuracy
- Alternative solutions

DOI: 10.1201/9781003293453-3

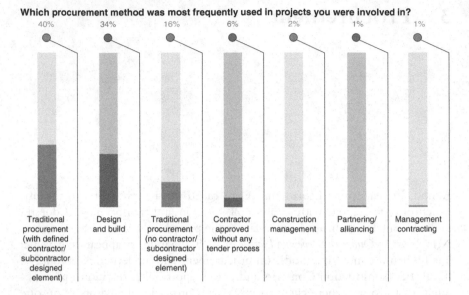

Figure 3.1 Frequently used procurement.

Source: The RIBA Construction Contracts and Law Report 2022.

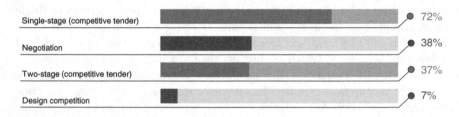

Figure 3.2 Tendering methods.

Source: The RIBA Construction Contracts and Law Report 2022.

- Functional performance built to an agreed timescale and quality
- Value for money
- Quantified levels of risk
- No surprises
- Legal compliance
- End-user satisfaction
- No reputational damage.

Source: CLC | Procuring for Value

These are outcomes which clients can, and should, influence and control by using their buying power and by defining what value means to them as a client or asset owner.

BIS Research paper No.145 Supply Chain Analysis into the Construction Industry concluded that for a typical £10 million project a tier 1 contractor may be directly managing around 40 tier two sub-contractors. With typically 50% to 75% of the value of the construction work accounted for by 4 or 5 large, complex tier 2 sub-contracts. In addition, complex tier 2 sub-contracts are also highly disaggregated, typically featuring over 30 suppliers at tier 3. The paper went on to conclude; *The current structure facilitates cost reduction through downward competitive pressure, which may not secure best value delivery through the coordination of activities on site.*

Not withstanding the above the study identified a number of crucial factors which determined successful project delivery as follows:

- The availability of equitable finance
- Certainty of early payment
- Early contractor engagement
- Continuing involvement of the supply chain in design development
- Strong relations
- Collaboration with suppliers.

The study concluded that construction has a low awareness of waste and duplication that are embedded in current procurement practice.

New models of procurement

The Cabinet Office reviews its approach to construction procurement at regular intervals. In July 2014, the Cabinet Office recommended three models of procurement in the aptly named New Models of Procurement. They are:

- Cost Led Procurement
- Integrated Project Insurance
- Two Stage Open Book.

The common principles of these models are a requirement for clients to do the following:

- Clearly define the desired functional outcome including specific require-ments, for example, carbon reduction and use of apprentices
- Identify typical costs and delivering the outcomes based on available data, benchmarking and cost-planning work. This will enable the client to set a realistic yet challenging cost ceiling, that would be achieved or bettered, and costs would be further reduced over a series of projects or programmes of work
- Engage with the supply chain that embraces the principles of Early Contractor Involvement and a high level of supply chain integration; and ensure that on completion of the capital phase the specified output performance is achieved

- Apply a robust review process to ensure appropriate scheme definition, create commercial tension, monitor scheme development and address any unnecessary scope, risks and potentially missed opportunities
- Take steps to ensure that those appointed to carry out the processes of the models, whether internal or external to the client organisation have the skills to do so effectively.

The hope is that these models will change the way in which Government buys construction. They attempt to change the procurement process so that the supply chain responds to an outline declared budget and client requirements. This is different from the usual process where the price is built up against a specification without the tendering organisations knowing what budget the client has available. It is claimed that pilot projects delivered savings of between 6% and 20% (The annual report of the NMCP Working Group to the IPA, January 2019).

Cost-led procurement

The client, who may have existing framework agreements, invites one or more integrated teams or supply chains to deliver a project using collaborative behaviour, tools and techniques under the declared budget on the initial project and then is required to achieve cost savings on future projects for the same quality standard. In competition with other similar integrated supply chain organisations from the framework, two to three organisations are involved early in the procurement process to encourage the use of innovative design and working practice to develop bids that identify cost reductions. To be successful, the bidding organisations must achieve a price below the budget and are scored on a mix of commercial and physical benefits. If there are no prices below the budget, then others from outside of the framework are permitted to bid. If the budget cannot be met or improved upon, then the client will have to consider increasing the budget or reducing the specification.

Integrated project insurance (IPI)

The origins of the Integrated Project Insurance (IPI) go back over a decade, as it was being discussed in the early 2000s when I was part of the team that developed the Strategic Forum for Construction's 'Integration Tool Kit' for developing integrated project teams and supply chains. In Europe, decennial liability insurance has been in the marketplace for many years.

To implement IPI, the client holds a competitive appointment process for selecting an IPT who will be responsible for project delivery. Criteria for selecting the IPT will include the common requirements for capability – such as proven track record – plus demonstration of maturity behaviours, lean thinking, removal of wasteful processes and fee. The successful team then creates a preferred solution that will provide the outcome required by the client, including the generation of savings measured against historical cost benchmarks. The differentiator between this and other forms of procurement is the insurance policy which covers all the usual construction insurances for the client and all supply chain members involved in the project.

The insurance also absorbs some of the commercial risk of cost escalation beyond the budget. However, before that facility is available, there is a pain-share threshold divided up on an open basis between the client and the other supply chain members.

This IPI model includes independent facilitation and a gateway process which ensures value for money and helps balance commercial risk to a level at which an insurer is prepared to provide the insurance. This third-party intervention and review process is seen as a considerable benefit to the successful outcome of the project. The insurance provides for cost overruns only up to a financial cap, thus limiting the insurer's potential exposure and will not meet all of an overspend but helps remove some of the blame culture prevalent within the industry. Payment under the policy will be based on proven loss. However, to obtain the insurance cover at the outset, the proposed scheme will be subjected to independent validation to ensure that the project is deliverable and that it is commercially competitive or, as the guidance refers to it, as the maintenance of 'commercial tension'.

The first UK construction contract to use integrated project insurance was awarded in 2015 for the £10 million Centre for Advanced Building Technologies at Dudley College in the West Midlands which was completed in 2017. All the project team firms signed up to a new alliancing contract, CABTech, which also features a project bank account under which the firms were jointly liable for the project.

Two-stage open book

This is a two-stage tender process where a contractor is selected after the first phase rather than at the end of a two-stage tendering process. The first stage involves contractors and their consultant team being selected based on 'capacity, capability, stability, experience, strength of their supply chain and fee (profit plus company overhead)'. The second stage consists of the successful team being asked to produce a proposal on an open book basis against the client's brief and cost benchmarks. The intention is that this will allow early engagement of the supply chain leading to an agreed price and risk profile prior to the client committing to construction. An interesting feature of this approach is that it is intended that the client will be able to 're-engineer' the supply chain as part of a joint review, with the potential of tier 1 contractors sharing their supply chain members with the client and other tier 1 contractors.

Two of the models really are further developments of existing procurement methods. But the cost-led procurement method is a departure from previous practice in that the client is now going to declare the budget openly and up front. If the bidding contractors cannot meet the budget, then the project does not go ahead unless more finance is found, or the specification level is reduced. The two-stage open book method is really about the client getting access to the supply chain early in the procurement process and ring fencing the overhead and profit fee at stage 1. In reality what does a main contractor (tier 1 contractor) do other than take risk, provide management resource and expertise and broker the contractual relationship between client, itself, the subcontractors and suppliers?

But what about construction management? Could the public sector 'go direct' and manage its own supply chains? It would appear not, although surely it is food for thought given that the client teams will need to have competent project and cost management capability to procure these models. The real game changer and innovation, as far as the UK construction industry is concerned, is the IPI. It has been a long time coming but it is about to surface both within the public sector and private sector. The issue will be how much will it cost in terms of premium and whether it is affordable in terms of the cost of the premium when weighed up against the project risks.

The Construction Playbook

Perhaps the most significant report to be issued by the UK Government since the fourth edition of *New Aspects of Quantity Surveying Practice* was *The Construction Playbook* in December 2020, to be implemented by the end of 2023. The playbook will apply to all new public works projects and programmes undertaken by Central Government departments and arm's length bodies (ALB) on a *'comply or explain basis'*. The term 'playbook' has been adopted from American football coaching to describe all the plays and strategies available, leaving it up to the team to select the most appropriate. Aimed primarily at those within the public sector responsible for the planning, procurement and delivery of projects and programmes, it is hoped that the processes in the playbook will, in time, trickle down to the private sector.

Currently, main contractors have positioned themselves into a place where they are unable to make money by using the existing broken and unsustainable business models and unfortunately the industry is unable to solve this problem due to lack of resources. According to the Construction Financial Management Association (www.cfma.org), the average pre-tax net profit for general contractors is between 1.4% and 2.4% and for subcontractors between 2.2% and 3.5%. The Construction Playbook aims to rebalance the relationship between clients and contractors by moving away from selection purely on lowest price and trying to ensure that contractors and sub-contractors maintain sustainable profit margins. There is also strong emphasis on early supply chain involvement. The Construction Playbook has been structured around the main stages of a typical procurement and project lifecycle:

• Preparation and planning
• Publication (Tendering)

- Selection
- Evaluation and award
- Contract implementation.

However, for those hoping to find a silver bullet in the playbook, there will be disappointment as it proposes little that has not been proposed previously. What is new is that previous initiatives are now incorporated into a single document. The playbook has its roots in the Carillion failure and subsequent supply chain fall out and Boris Johnson's pledge to deliver *'better, greener faster'* public works and is based on the perceived success of the Outsourcing Playbook now the Sourcing Playbook. The Construction Playbook aims to:

- Adopt outcome-based specifications designed with the input of industry to ensure continuous improvement and innovation
- Where appropriate seek longer term contracting across portfolios instead of project by project
- Standardise designs, components and interfaces as far as practical
- Drive innovation and Modern Methods of Construction, through standardisation and aggregation of demand, increased client capability and setting clear requirements of suppliers
- Create sustainable, win–win contracting arrangements that incentivise better outcomes, improve risk management and promote the general financial health of the sector
- Strengthen financial assessment of suppliers
- Increase the speed of end-to-end project and programme delivery.

The Construction Playbook has 14 key policies, 8 have been adopted from the Outsourcing Playbook with another 6 policies new to the Construction Playbook as discussed below;

1 *Commercial pipelines*

Informing suppliers of Government's proposals for up to five years ahead, enabling better preparation and investment using longer-term contracts although in practice, this seems to favour tier one contractors and marginalise many SMEs.

2 *Market health & capability assessment*

BREXIT, an aging workforce and skills shortages impact on the ability to deliver a project. The state of the market and its capability to meet the requirements of a programme or project should be assessed up front to identify limitations and possible new technologies.

3   *Portfolios and longer-term contracting*

Linked to Playbook Policy 1, Commercial pipelines, this policy advocates a move away from short-termism to a long-term, almost a partnering approach. The Government has always been a fan of Modern Methods of Construction (MMC) and not for the first time in the Playbook a move to MMC and a manufacturing approach in construction is given heavy emphasis claiming it will bring improved productivity and efficiency savings.

4   *Harmonise, digitise and rationalise demand*

The industry in tasked to seek opportunities to develop and adopt common standards and specifications and use cross-sector product platforms to create a more resilient pipeline and drive efficiencies, innovation and productivity. Once again emphasis is placed on public sector procurement to support investment in MMC and seems to envisage a situation where off site/standardised manufactured components form the basis of projects.

During the procurement process, the potential for the use of MMC should be explored with the benefits of off-site factory production in the construction industry are well documented and include the potential to considerably reduce waste especially when factory manufactured elements and components are used extensively. Its application also has the potential to significantly change operations on site, reducing the number of trades and site activities and changing the construction process into one of a rapid assembly of parts that can provide many environmental, commercial and social benefits, including Benefits of Modern Methods of Construction in Housing:

- Reduced construction programme cost (20%–40%) time (20%–60%) and improved quality
- 70% less on-site labour, improved Health and Safety and local employment opportunities
- Fewer deliveries to site and more efficient materials use
- Opportunities for customisation
- 20%–33% lower energy in use
- Lighter-weight construction.

Demand for housing in the UK is growing with approximately 3.9 million new homes required. The Government aims to develop 300,000 new homes a year to meet this demand. There is a general recognition that construction targets cannot be met without extensive use of MMCs. However, only 15,000 homes per year are currently factory-made and the reasons for slow uptake of MMC (UK Govt 2019) include:

- The fact that many MMC-based solutions are relatively untested. Housing providers need evidence on in-use performance, maintenance and repair

costs, adaptability, resilience (e.g., addressing concerns of warranty/mortgage providers), and compliance with regulations to support selection
- Perceived risks associated with of limited supply chain capacity, poor productivity and reliability in delivering homes that meet quality standards
- Poor public perception
- The impact of skills shortages.

Off-site construction is one of a group of approaches to more efficient construction that also include prefabrication, improved supply chain management and other approaches. Technologies used for off-site manufacture and prefabrication range from modern timber and light gauge steel framing systems, tunnel form concrete casting through to modular and volumetric forms of construction and offer great potential for improvements to the efficiency and effectiveness of UK construction. Whether this blind faith in MMC is justified remains to be seen, remember a similar push for system building in the 1960s?

5 *Further embed digital technologies*

The penetration that BIM has made into construction practices since it was mandated by the UK Government in 2016 has been discussed elsewhere in this book. No surprise there fore that the Construction Playbook encourages the use of the UK BIM Framework.

6 *Early supply chain involvements*

The suggestion is to involve the supply chain early in the procurement process and to establish a 'mutually beneficial, open and collaborative approach'. Getting tier 2 and 3 sub-contractors involved in the procurement process is obviously a good way towards reaching 'baked in buildability', the problem historically is ensuring that early input is compensated.

7 *Outcome-based approach*

The use of performance-based specifications is advocated focused on whole life value, performance and cost, without being prescriptive about how to deliver outcomes. In theory, this approach unlocks innovation and drives continuous improvement but can be abused. A new project scorecard is being developed by the Infrastructure and Project Authority to support this policy.

8 *Benchmarking and 'Should Cost Models'*

Using benchmarking of key project deliverables and outcomes to drive consistency and robustness of cost estimates. The aim being to generate the inputs required for Should Cost Models (SCMs, do we really need another term?) facilitating evaluation of whole life cost and risk.

A Should Cost Model (SCM) provides a forecast of what a service, project or programme 'should' cost over its whole life. For public works projects, SCMs forecast costs over a period that includes both the build phase and the expected design life. This includes costs of additional market factors such as risk and profit. It provides an understanding of whole life costs, including the impact of risk and uncertainty on both cost and schedule. Notably, the key factor is 'whole life cost' and not the initial purchase price. SCMs should be used early in the procurement process.

9   *Delivery model assessments*

Follow an analytical, outcome-based process early in the preparation/planning stages of a project to decide the most appropriate delivery model, enabling clients and industry to work together to deliver the best possible outcomes by determining the optimal split of roles and responsibilities by considering the following:

- Set up an appropriate cross-functional team and identify key stakeholders. Agree the sponsor and governance approach including project board. Define the desired outcomes for the project and set these out in a project scorecard
- Identify data inputs and potential delivery model approaches
- Consider the strategic and operational approach
- Strategy and supplier interaction
- Consider how to maximise the use of innovations, digital solutions and MMC and the market's capability to deliver this. What level of involvement is required in specifying design output?
- Consider the required internal capability and capacity (including senior management and supporting functions). Consider the approach to any potential asset ownership including any new intellectual property rights
- Identify the risks that may impact the value profile. Who is best placed to manage these risks and what impact would this have on where activities sit?
- Assess the whole life cost of the project
- Recommendations.

10   *Effective contracting*

Selecting an appropriate standard form construction contract with appropriate KPIs and boilerplate clauses to suit the type and complexity of works, intended outcomes, delivery model, procurement strategy and commercial approach. The aim being to support an exchange of data, drive collaboration, improve value and manage risk. The Playbook expects a fully integrated, consistent suite of documents. Frameworks will be reviewed with a view to consolidating, where appropriate, and adopting a new 'gold standard' for frameworks.

11   *Risk Allocation*

Ensure that risks are analysed and allocated to, and managed by, those best able to manage them to deliver value for money and successful outcomes. The fundamental principle being that contracts are profitable and offer a fair return to ensure the market is sustainable. Risk management is discussed in Chapter 6.

12   *Payment mechanism and pricing approach*

As a general principle, the approach should be to link payment to the delivery of outputs and/or the work value and supplier performance to ensure it incentivises the desired behaviours or outcomes. Contracts should be designed to be profitable and offer a fair return for the market to be sustainable in the long term.

13   *Assessing the economic and financial standing of suppliers*

During the selection process, clients should assess the economic and financial standing of suppliers against the minimum standard to safeguard the delivery of projects against a supplier going out of business during the life of the contract. The Government now expects these assessments to be carried out in all construction procurements, as for other services. The key being that they are tailored to individual projects, and are proportionate, fair and transparent.

14   *Resolution planning*

Following the fall out of the collapse of Carillion, all new critical public works contracts will now require resolution planning information to be provided by suppliers to help ensure the continuity of projects and their orderly transfer to a new supplier in the event of supplier insolvency.

Cross-cutting the 14 key policies and embedded in project planning are the priorities of

- Health and safety/wellbeing and addressing that lack of appeal that construction has for new entrants
- Building safety, avoiding more Grenfell tower disasters
- Building back greener – meeting UK Government targets of net zero by 2050.

ISO 20400 – Standard for Sustainable Procurement

This standard provides guidance for organisations of all sizes and nationalities wishing to implement a responsible resourcing strategy. The document advises how best practice and value may be achieved, while respecting ethical standards, promoting diversity and equality, and minimising negative impacts and demands on resources. The guidance, supersedes BS 8903 and covers:

- Life cycle considerations
- Due diligence
- Risk management
- Ethical and social responsibility
- Corporate social responsibility.

The arrangement of the new international standard, provides systematic, step-by-step guidance from strategy through to implementation in the following areas:

- *Fundamentals* – the definitions, concepts, drivers and principles of sustainable procurement
- *Policy and strategy* – key issues for organisations to consider when developing a sustainable procurement policy and strategy
- *Organising procurement* – creating the organisational conditions necessary to procure sustainably and setting priorities
- *Procurement process* – how to procure differently to best practice guidelines.

International Standards Organisation (ISO) 20400 is a guidance document only and so cannot be certified as with other ISO requirement standards. It provides a benchmark for firms to determine how sustainable their procurement practices are, what improvements are required and how they may be implemented. Organisations are given guidance on how to make the right decisions, assign responsibility, achieve supply chain security and reduce supply chain risks. Despite the attraction for clients of adopting sustainable construction, as discussed in Chapter 2, sustainability and green issues remain a nebulous topic for many within the construction industry. Popular perception is that there is a lack of customer demand for sustainability to be considered during design and procurement stages, however, consider the following reasons to be green:

- The Dept. for Environment, Food and Rural Affairs requires developers to complete a Sustainable Development Impact Test
- Public sector contractors must achieve BREEAM excellence for all new buildings
- Many high-profile private developers and landowners are seeking the same or higher standards of sustainability performance from their partners
- Investors are becoming increasingly interested in sustainability and are encouraging property industry partners to do the same.

The measures adopted to assess sustainability performance, developers and design teams are encouraged to consider these issues at the earliest possible opportunity, are:

BREEAM (Building Research Establishment Assessment Method)

BREEAM-In-Use, for existing buildings

New National Technical Standards.

BREEAM

BREEAM has been developed to assess the environmental performance of both new and existing buildings. BREEAM assesses the performance of buildings in the following areas:

- Management: overall management policy, commissioning and procedural issues
- Energy use
- Health and wellbeing
- Pollution
- Innovation
- Transport
- Land use
- Ecology
- Materials
- Water, consumption and efficiency.

In addition, unlike the National Technical Standards, BREEAM covers a range of building types such as:

- Offices
- Industrial units
- Retail units
- Schools
- Other building types such as leisure centres can be assessed on ad hoc basis.

In the case of an office development the assessment would take place at the following stages:

- Design and procurement
- Management and operation
- Post construction reviews and
- Building performance assessments.

A BREEAM rating assessment comes at a price and according to the fee scale for BREEAM assessors to carry out an assessment at each of the above stages could be between £800 and £4,160, an item that should be considered when completing Section 5 of the RICS New Rules of Measurement 1; Order of Estimating and Cost Planning.

Some of the benefits of having a BREEAM rating are claimed to be:

Demonstrating compliance with environmental requirements

Marketing; as a selling point to tenants and customers

Financial; to achieve higher and increased building efficiency.

A number of key issues were relating to BREEAM are as follows:

- **Timing:** many BREEAM credits are affected by basic building form and servicing solutions. Cost effective BREEAM compliance can only be achieved if careful and early consideration is given to BREEAM related design and specification details. Clear communication between the client, design team members and in particular, the project cost consultants, is essential
- **Location:** building location and site conditions have a major impact on the costs associated with achieving very good and excellent compliance
- **Procurement route:** PPP and similar procurement strategies that promote long-term interest in building operations for the developer/contractor typically have a position influence on the building's environmental performance and any costs associated with achieving higher BREEAM ratings.

Part L of the Building Regulations were revised in 2021 and introduced tougher energy and environmental section, these new regulations are mandatory from June 2022. In addition, the long awaited 2021 Environment Act allows the UK to enshrine some environmental protection into law.

Consequently, new energy performance standards for buildings and large existing buildings are required and quantity surveyors must be able to provide cost advice on alternative solutions. So how does BREEAM work? BREEAM measures the environmental performance of buildings by awarding credits for achieving a range of environmental standards and levels of performance. Each credit being weighted according to its importance and the resulting points are added up to give a total BREEAM score and rating (see Figure 3.3). Additional weights are applied to other building categories namely; Simple buildings, Shell and core only and Shell only.

BREEAM is assessed over several categories (see Figure 3.4). Each category contributes a percentage towards the overall rating.

The higher the BREEAM rating the more mandatory requirements there are and progressively harder they become. In 2011, BS ISO 15686-5 Service

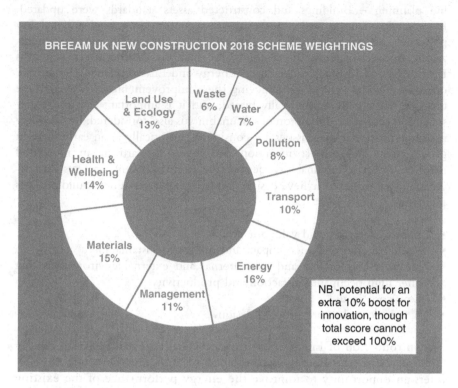

Figure 3.3 BREEAM scores.

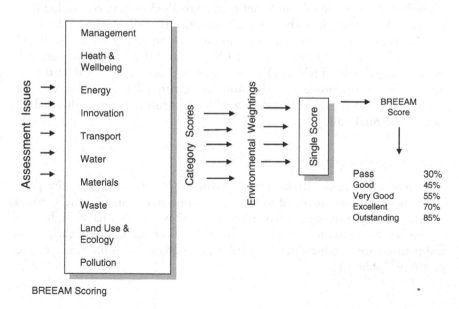

Figure 3.4 BREEAM scoring.

life planning – Buildings and constructed assets standards were updated. Increasingly clients as well as end users are requesting improved sustainability performance from their buildings, over and above the regulatory requirements arising from changes in the Building Regulations. Methodologies such as BREEAM and LEED (Leadership in Energy and Environmental Design) are often used as the vehicle for achieving these improvements. However, these tools are largely environmentally biased, and it is important that the wider social and economic dimension of sustainability is also considered. It is strongly recommended that these issues are considered holistically at an early stage in procurement and project inception and taken forward in an integrated manner. From a sustainability perspective, refurbishment projects are increasingly expected to achieve design standards expected of new build projects including:

- Improved quality and value for money
- Reduced environmental impact and improved sustainability
- Healthy, comfortable and safe internal and external environments that offer high occupant satisfaction and productivity
- Low costs in use
- A flexible and future-proofed design.

The introduction of Energy Performance Certificates and Display Energy Certificates on the Energy Performance of Buildings Directive (EPBD) offers an opportunity to improve the energy performance of the existing building stock and setting out on a refurbishment or refit without ensuring that an improvement of energy performance is specified would be ill-advised as there are significant benefits to be realised both in cost reductions as well as reductions in carbon dioxide emissions. Perhaps one of the most informative pieces of research into the cost of complying with BREEAM was carried out jointly between Cyril Sweett and the BRE. The research tried to dispel the widely held view that improving BREEAM ratings is necessarily an expensive exercise and demonstrated that an increase of between 1 and 3 BREEAM rating levels can be achieved at an additional up to 2% of capital cost.

Measuring and costing sustainability

The elements with the highest maintenance to capital cost ratios are generally those that are included in major refurbishments, mainly the services elements and the fittings. Conversely external works also has a high ratio because of the constant repletion of small items of work, particularly grass and planted areas, which need constant maintenance at some times of the year (see Table 3.1).

Table 3.1 Elements with the highest maintenance to capital ratio

Element	Maintenance* as % of Capital
Fittings	210%
External Works	148%
Heating	133%
Electrical Installation	118%
Wall finishes	92%

Note
* NPV over 60 years at 3.5% discount rate.

What items should be considered by a quantity surveyor?

Procurement is the whole process of acquisition from third parties covering goods services and capital projects. The process spans the whole life cycle from initial concept through to the end of the useful life of the asset (including disposal) or end of the services contract.

Sustainable procurement is a key method for delivering an organisation's sustainable development priority. It is all about taking social and environmental factors into consideration alongside financial factors in making these decisions. It involves looking beyond the traditional economic parameters and making decisions based on the life cycle cost, the associated risks, measures of success and implications for society and the environment. Making decisions in this way requires setting procurement into the broader strategic context including value for money, performance management, corporate and community priorities.

In general terms, the quantity surveyor should consider the implications of ethical procurement that encompasses:

1 Whole life costs/life cycle costs
2 Costs of alternative (renewable) materials
3 Renewable energy schemes
4 Recycled contents schemes and
5 The ethical sourcing of materials and labour.

Of course, the majority of building sock, approximately 98%, is existing and therefore one of the biggest challenges for the construction industry is how to deal with this stock. For example, approximately one-third of total CO_2 emissions is from commercial buildings of which commercial offices is a major contributor. The actions of both landlords and tenants contribute to building energy performance and in turn the CO_2 emissions produced. The landlord has sole control over the building fabric performance whereas the tenant is responsible for using the building and controls hours of use, IT equipment and management of the setting of temperatures. As part a IPF Research Programme a group lead by Cyril Sweett investigated the cost of making energy efficient improvements to existing 1990 built commercial buildings held in investment portfolios. The research found that in the case of a 1990-built deep plan air-conditioned office building that for a base line cost of £1000/m² reductions of 25% could be achieved by modernising older type buildings.

Whole life costs/life cycle costs

Although life cycle costing can be carried out at any stage of the project and not just during the procurement process, the potential of its greatest effectiveness is during procurement because:

- Almost all options are open to consideration at this time
- The ability to influence cost decreases continually as the project progresses, from 100% at project sanction to 20% or less by the time construction starts
- The decision to own a building normally commits the user to most of the total cost of ownership and consequently there is a very slim chance to change the total cost of ownership once the building is delivered.

Typically, about 75%–95% of the cost of maintaining and repairing a building is determined during the procurement stage. The service life of an element – product or whole building may be viewed in one or more of the following ways:

- Technical life – based on physical durability and reliability properties
- Economic life – based on physical durability and reliability properties
- Obsolescence – based on factors other than time or use patterns, for example, fashion.

Common terms used to describe the consideration of all the costs associated with a built asset throughout its life span are:

- Costs-in-use
- Life cycle costs
- Whole life costs
- Through life costs and other.

Definitions given in the BS ISO 15685:5 are as follows:

- *Life cycle costing* ... is a methodology for the systematic economic evaluation of the life cycle costs over the period of analysis, as defined in the agreed scope
- *Whole life costing* ... is a methodology for the systematic economic consideration of all the whole life costs and benefits over the period of analysis, as defined in the agreed scope.

The ISO defines lifecycle costs as 'the cost of an asset throughout its lifecycle while fulfilling the performance requirements'. Lifecycle costing is basically a simple concept – it answers the question: 'If I build this building, what future costs will I be letting myself in for?' So, it is only a projection of the costs that result from commissioning a building, and which will be the responsibility of the client. While life cycle costs are not difficult, it is complex because potentially there are a huge number of costs to consider and a wide range of variables. It is also complicated by the introduction of time into the equation and therefore the ways of how to treat the effects of inflation, and lost investment opportunities or money. The chief use of life cycle cost is, at the design stage, to compare the relative cost of alternative materials or forms of construction. It also enables the client/building owner to appreciate the possible disruption caused by repairs and maintenance to the alternatives under consideration.

Part 5 of ISO 15685:5 (2017) covers life cycle costing as follows:

- Definitions, terminology and abbreviations
- Principles of life cycle costing – that is, Purpose and scope; What costs to include/exclude; Typical forms and level of LCC analysis at key stages; Outputs
- Forms of LCC calculations and six methods of economic evaluation (with informative examples)
- Setting the scope for LCC studies including how to deal with risks and uncertainty
- How LCC forms part of the whole life costing business investment option appraisal process
- Reporting and analysis techniques.

The ISO allows organisations to build up life cycle costs of construction projects on a common basis. At present, there is no way to compare LCC estimates, and few organisations are able to estimate LCC. The ISO helps to eliminate confusion in the industry and is likely to become the established methodology going forward.

First introduced to the UK construction industry over six decades ago by Dr P A Stone as costs-in-use; it is only recently, with the widespread adoption of Public Private Partnerships in the early 2000s as the preferred method of procurement by most public sector agencies, that the construction industry has started to see some merit in life cycle costing. In addition, building owners with long term interests in property are starting to demand evidence of the future costs of ownership. For example, PPP projects are commonly awarded to a consortium based on Design, Build and Operate (DBO), and contain the provision that, at the end of the concession period, typically 25 years, the facility is handed back in a well-maintained and serviceable condition. This is of course in addition to the operational and maintenance costs that will have been borne by the consortia over the contract period. Therefore, for PPP projects given the obligations touched on above, it is clearly in the consortiums' interest to give rigorous attention to costs incurred during the proposed assets life cycle to minimise operational risk. Although Stone's work was well-received in academic circles, where today extensive research continues in this field, there has been and continues to be, a good deal of cynicism in the UK construction industry to the wider consideration of life cycle costs.

It has been estimated that the relationship between capital costs: running costs and business costs in owning a typical office block over a 30-year period is:

Construction (capital) cost	1
Maintenance and operating costs	5
Business operating costs	200

Source: The long-term costs of owning and using buildings –
Royal Academy of Engineering and Stanhope (1998).

Some doubt has been expressed on the accuracy of the Royal Academy ratio and that it was not research based, *Exposing the myth of the 1:5:200 ratio relating initial cost, maintenance and staffing costs of office buildings*. Research carried out by the University of Reading put the ratio closer to 1:0.4:12, the fact remains that whole life costs are still a considerable factor in the cost of built asset ownership.

The business operating costs in the above equation include the salaries paid to staff, etc. Clearly in the long term, this aspect is worthy of close attention by design teams and cost advisors. After all, it can cost up to £30,000 to replace a staff member who leaves an organization due to an unsatisfactory working environment. One of the reasons for this lack of interest is, particularly in the private sector and the developer's market that, during the 1980s, financial institutions became less enamored with property as an investment and turned their attention to the stock market. This move led to the emergence of the developer/trader. Often an individual, rather than a company, who proposed debt finance rather than investment financed development schemes. Whereas previously, development schemes had usually been pre-let and the investor may even have been the end user, the developer trader had as many projects in the pipeline as finance could be obtained for. The result was an almost complete disregard for life cycle costs as pressure was put on the designers to pare capital costs at the expense of ultimate performance as building performance is poorly reflected in rents and value. Fortunately, these sorts of deals have all but disappeared, with a return to the practice of pre-letting and a very different attitude to life cycle costs. For if a developer/trader was developing a building to sell on they would have little concerned with the running costs, etc. however, in order to pre let a building, tenants must be certain that, particularly if they are entering a lease with a full repair and maintenance provision, that there are no 'black holes', in the form of large repair bills, waiting in the devour large sums of money at the end of the lease. In the present market therefore, sustainability is as important to the developer as the owner/occupier. A building will have a better chance of attracting better quality tenants, throughout its life, if it has been designed using performance requirements across all asset levels, from Facility (Building), through System (Heating and Cooling System), to Component (Air Handling Unit), and even Sub-Component (Fans or Pumps).

In and around major cities today, buildings that attracted good tenants and high rents in the 1980s and early 1990s are now tending to only attract secondary or tertiary covenants, in multiple occupancies, leading to lower rents and valuations. This is an example of how long-term funders are seeing their 25–35 year investments substantially underperforming in mid-life, thus driving the need for better whole life procured buildings.

Life cycle cost procurement includes the consideration of the following factors:

- **Initial** or procurement costs, including design, construction or installation, purchase or leasing, fees and charges
- **Future** cost of operation, maintenance and repairs, including: management costs such as cleaning, energy costs, although energy costs are often excluded as the price volatility in this sector adds yet another unknown to the study. For example, the increase in the cost of wholesale gas following the Russian invasion of Ukraine in 2022
- **Future** replacement costs including loss of revenue due to non-availability
- **Future** alteration and adaptation costs including ditto
- **Future** demolition/recycling costs.

Life cycle cost appraisers may include whatever they deem to be appropriate – provided they observe consistency in any cross-comparisons. The timing of the future costs associated with various alternatives must be decided and then using a number of techniques described below asset their impact. Classically, life cycle cost procurement is used to determine whether the choice of say a component, with a higher initial cost than other like for like alternatives is justified by being offset by reduction of the future costs as listed above. This situation may occur in new build or refurbishment projects. In addition, life cycle cost procurement can be used to analyse whether in the case of an existing building a proposed change is cost effective when compared against the 'do nothing' alternative. There are three principal methods of evaluating whole life costs:

- Simple aggregation
- Net present value
- Annual equivalent.

Simple aggregation

The basis of life cycle costs is that components or forms of construction that have high initial costs will, over the expected life span, prove to be cheaper and hence better value than cheaper alternatives. This method of appraisal involves adding together the costs, without discounting, of initial capital costs, repair and maintenance costs. This approach has a place in the marketing brochure and it helps to illustrate the importance of considering all the costs associated with a particular element but has little value in cost forecasting. A similarly simplistic approach is to evaluate a component on the time required to pay back the investment in a better-quality product. For example, a number of energy saving devises are available for lift installations, a choice is made based on which over the life cycle of the lift, say five or ten years will pay back the investment most quickly. This last approach does have some merit, particularly in situations where the life cycle of the component is relatively short and the advances in technology and hence the introduction of a new and more efficient product likely.

Net present value approach

The technique of discounting allows the current prices of materials to be adjusted to take account of the value of money during the life cycle of the product. Discounting is required to adjust the value of costs, or indeed, benefits which occur in different time periods so that they can be assessed at a single point in time. This technique is widely used in the public and the private sectors as well as sectors other than construction. The choice of the discount rate is critical and can be problematic as it can alter the outcome of calculation substantially. However, when faced with this problem, the two golden rules that apply are; that in the public sector follow the recommendations of the Green Book or Appraisal & Evaluation in Central Government 2022, currently recommending a rate of 3.5% for long-term discounting. In the private sector, the rule is to select a rate that reflects the real return currently being achieved on investments. To help in understanding the discount rate, it can be considered almost as the rate of return required by the investor which includes costs, risks and lost opportunities.

The mathematical expression used to calculate discounted present values are set out as follows:

$$\text{Present Value (PV)} = \frac{1}{(1 + i)^n}$$

where: (i) = rate of interest expected or discount rate
       (n) = the number of years.

This present value multiplier/factor is used to evaluate the present value of sums, such as replacement costs that are anticipated or planned at say 10 or 15-year intervals.

For example, consider the value of a payment of £150 that is promised to be made in five years' time.

Assuming a discount rate of 3.5%, £150 in five years' time would have a present worth or value of £126.30.

$$\frac{1}{(1 + i)^n}$$

$$£150 \times \frac{1}{(1.035)^n} = £150 \times 0.8420 = £126.30$$

Or in other words, if £126.30 were to be invested today @ 3.5%, this sum would be worth £150.00 in five years' time, ignoring the effects of taxation.

Calculating the present value of the differences between streams of costs and benefits provides the net present value (NPV) of an option and this is used as the basis of comparison as follows;

Annual equivalent approach

This approach is closely aligned to the theory of opportunity costs and is used as a basis for comparison between alternatives, i.e., the amount of interest lost by choosing option A or B as opposed to investing the sum at a given rate percent. This approach also can include the provision of a sinking fund in the calculation in order that the costs of replacement are also provided for.

In using the annual equivalent approach, the following equation applies:

Present Value of £1 per annum, (sometimes referred to by actuaries as the Annuity that £1 will purchase).

This multiplier/factor is used to evaluate the present value of sums, such as repair and maintenance costs that are paid on a regular annual basis

$$\text{Present Value of } £1 \text{ per annum} = \frac{(1 + i)^n - 1}{i(1 + i)^n}$$

where: (i) = rate of interest expected or discount rate
        (n) = the number of years

Previously calculated figures for both multipliers are readily available for use from publications such as Parry's Valuation Tables.

Sinking funds should also be considered; a fund created for the future cost of dilapidations and renewals. Given that systems are going to wear out or need partial replacement, it is thought to be prudent to 'save for the rainy day' by investing capital in a sinking fund to meet the cost of repairs, etc. The sinking fund allowance therefore becomes a further cost to be considered during the evaluation process. Whether this approach is adopted will depend on a number of features including, corporate policy, interest rates, etc.

Life cycle costing is not an exact science, as, in addition to the difficulties inherent in future cost planning, there are larger issues at stake. It is not just a case of asking 'how much will this building cost me for the next 50 years', rather it is more difficult to know whether a particular building will be required in 50 years' time at all – especially as the current business horizon for many organisations may be closer to 10 years. Also, life cycle costing requires a different way of thinking about cash, assets and cash flow. The traditional capital cost focus has to be altered, and costs be thought of in terms of capital and revenue costs coming from the same 'pot'. Many organisations are simply not geared up for this adjustment. The common misconception that a life cycle costed project will always be a project with higher capital costs does not assist this situation. As building services carries a high proportion of the capital cost of most construction projects, this is of particular importance. Just as capital and revenue costs are intrinsically linked so are all the variables in the financial assessment process. Concentrate on one to the detriment of the others and you are likely to fail.

Perhaps, the most crucial reason is the difficulty in obtaining the appropriate level of information and data to make the calculations reliable. Clift and Bourke (1999) found that despite substantial amounts of research into the development of database structures to take account performance and LCC, there remains significant absence of standardisation across the construction industry in terms of scope and data available. Ashworth also points out that the forecasting of building life expectancies is a fundamental prerequisite for life cycle cost calculations, an operation that is fraught with problems. While to some extent building life relies on the lives of the individual building components, this may be less critical than a first imagined, since the major structural elements, such as the substructure and the frame, usually have a life far beyond those of the replaceable elements (see Chapter 2). Clients and users will have theoretical norms of total life spans, but these have often proved to be widely inaccurate in the past. The Building Maintenance Information of the Royal Institution of Chartered Surveyors was established in the 1970s. The BMI has developed a Standard Form of Property Occupancy Cost Analysis, which it is claimed allows comparisons between the cost of achieving various defined functions or maintaining defined elements. The BMI defines an element for occupancy cost as; expenditure on an item which fulfils a specific function irrespective of the use of the form of the building. The system is dependent on practitioners submitting relevant data for the benefit of others. The BMI has now been upgraded to the RICS Building Running Costs Online – see below. The increased complexity of construction means that it is far more difficult to predict the whole life cost of built assets. Moreover, if the mal function of components results in decreased yield or underperformance of the building then this is of concern to the end user/owner. There is no comprehensive risk analysis of building components available for practitioners, only a wide range of predictions of estimated life spans and notes on preventive maintenance – this is too simplistic, there is a need for costs to be tied to risk including the consequences of component failure. After all, the performance of a material or component can be affected by such diverse factors as:

- Quality of initial workmanship when installed on site and subsequent maintenance
- Maintenance regime/wear and tear. Buildings that are allowed to fall into disrepair prior to any routine maintenance being carried out will have a different life cycle profile to building that are regularly maintained from the outset
- Intelligence of the design and the suitability of the material/component for it usage. There is no guarantee that the selection of so-called high-quality materials will result in low life cycle costs.

RICS building running costs online

There have been various attempts to provide surveyors with information for LCC calculations. There are two major areas of concern:

- The parameters of the study
- The accuracy of the cost data.

The Building Running Costs Online is a subscription-based service run by the RICS and provides information and tools as follows:

- Building maintenance, decorations, fabric and operations, cleaning, utilities and administration costs for a wide range of building functions
- Figures for the life expectancy of over 300 common building components – substructure, superstructure, finishes, fittings and furnishings, and external works
- Compare the running costs of different building types, from a range of anything from one to 60 years
- Information on deterioration or failure of the components
- Calculation of the life cycle cost of a building
- Seven sets of indices that can be adjusted to UK regional location factors and dates
- A selection of the most common Wage Agreements and Dayworks Rates that are regularly updated and available online. Historical data back to 1987 is also available and shows the changes in labour costs over the last 20 years

- Average costs of refurbishment and rehabilitation by building type
- A 'calculator' which shows show the mean, median, and range of costs
- Twenty-seven hundred projects have been extracted from the BCIS database that include various levels of refurbishment. The average costs of refurbishment and rehabilitation can be searched by building type, and the 'calculator' will show the mean, median, and range of costs
- Subscriptions range from £735.00–£1,300.00 exc. VAT pa.

Other commonly voiced criticisms of whole life cycle costs are as follows:

Expenditure on running costs is 100% allowable revenue expense against liability for tax and as such is very valuable to property owners. There is also a lack of taxation incentive, in the form of tax breaks etc., for owners to install energy efficient systems.

In the short term and taking into account the effects of discounting the impact on future expenditure is much less significant in the development appraisal.

Another difficulty is the need to be able to forecast, a long way ahead in time, many factors such as life cycles, future repair and maintenance costs, and discount and inflation rates. LCC, by definition, deals with the future and the future is unknown. Increasingly obsolescence is being taken into account during procurement a factor that it is impossible to control since it is influenced by such things as; fashion, technological advances and innovation. An increasing challenge is to procure built assets with the flexibility to cope with changes. Thus, the treatment of uncertainty in information and data is crucial as uncertainty is endemic to LCC. Another major difficulty is that the LLC technique is expensive in terms of the time required. This difficulty becomes even clearer when it is required to undertake a LCC exercise within an integrated real-time environment at the design stage of projects.

In addition to the above, changes in the nature of development other factors have emerged to convince the industry that life cycle costs are important.

Life cycle cost procurement – critical success factors

- **Effective risk assessment** – what if this alternative form of construction is used?
- **Timing** – begin to assess LCC as early as possible in the procurement process
- **Disposal strategy** – is the asset to be owner occupied, sold or let?
- **Opportunity cost** – downtime
- **Maintenance strategy/frequency** – does one exist?
- **Suitability** – matching a client's corporate of individual strategy to procurement.

Other sources of cost data for life cycle cost calculations

In practice, most life cycle costs are associated with mechanical and electrical installations as these elements contribute, in many cases, a large percentage to overall costs. Choosing a realistic service life period for equipment is very important to ensure accurate valuation. A good source for service life expectancies is The Chartered Institution of Building Services Engineers (CIBSE) Guide M – Maintenance Engineering & Management Appendix 13.A1. NRM3 deals with the renewal and maintenance costs of buildings works and has aligned its structure to the Construction-Operations Building Information Exchange (COBie) data structure and definitions.

There now follows a simple example, based on the selection of material types, illustrating the net present value and the annual equivalent approaches to whole life cycle cost procurement (Table 3.2).

Table 3.2 Life cycle cost example

Material	Initial Cost	Installation Cost	Maintenance Cost per Day	Other Maintenance Costs	Life Expectancy
A	£275	£150	£3	£100 every 3 years for preservative treatment	12 Years
B	£340	£150	£3	None	15 Years

This problem is a classic one, which material, with different initial and maintenance costs will deliver the best value for money over the life cycle of the building. In this example, assuming a discount rate of 6%, it is assumed with an anticipated life of 25 years.

Table 3.3 indicates a life cycle cost calculation for material B presented in two ways; as a net present value and also as an annual equivalent cost. The calculation is repeated for each material or component under consideration and then a comparison can be made.

Clearly, the choice of the correct type of material or component would appear of critical importance to a client as future replacement and maintenance costs will have to be met out of future income. However, theory and practice are often very different. For example, for many public authorities, finding budgets for construction works is usually more difficult than meeting recurring running and maintenance costs that are usually included in annual budgets as a matter of course.

NRM3 Order of cost estimating and cost planning for building maintenance works

The final part of the NRM suite, NRM3, had a 'soft launch' in 2013 and came into effect from 1st January 2015. A second edition followed in October 2021

Table 3.3 Results for Material B

Total Discounted Costs

Year	Present Value of £1 per annum (PV of £1 pa)	Present Value (PV £1)	Initial Cost £	Other Costs £	Annual Cost £ £3 × 365	NPV of Replacement + Other + Annual Costs + Initial Costs £	Total NPV £	AEC £
1	0.943	0.943	490.00		1095.00	1523.02	1523.02	1614.4
2	1.834	0.890			1095.00	974.55	2497.57	1362.25
3	2.673	0.840			1095.00	919.38	3416.95	1278.31
4	3.465	0.792			095.00	897.34	4284.29	1236.41
5	4.212	0.747			095.00	818.25	5102.54	211.32
6	4.917	0.705			1095.00	771.93	5874.47	1194.65
7	5.582	0.665			095.00	728.24	6602.71	1182.76
8	6.210	0.627			1095.00	687.02	7289.72	1173.91
9	6.802	0.592			1095.00	648.13	7937.85	1167.04
10	7.360	0.558			1095.00	611.44	8549.30	1161.58
11	7.887	0.527			1095.00	576.83	9126.13	1157.13
12	8.384	0.497			1095.00	544.18	9670.31	1153.47
13	8.853	0.469			1095.00	513.38	10183.69	1150.35
14	9.295	0.442			1095.00	484.32	10668.00	1147.72
15	9.712	0.417		490.00	1095.00	661.37	11329.38	1166.50
16	10.106	0.394			1095.00	431.04	11760.42	1163.72
17	10.477	0.371			1095.00	406.64	12167.06	1161.28
18	10.828	0.350			1095.00	383.63	12550.69	1159.14
19	11.158	0.331			1095.00	361.91	12912.60	1157.24
20	11.470	0.312			1095.00	341.43	13254.02	1155.55
21	11.764	0.294			1095.00	322.10	13576.12	1154.03
22	12.042	0.278			1095.00	303.87	13879.99	1152.67
23	12.303	0.262			1095.00	286.67	14166.66	1151.45
24	12.550	0.247			1095.00	270.44	14237.10	1150.33
25	12.783	0.233			1095.00	255.13	14692.24	1149.36

Notes:

AEC = Annual equivalent cost.

Other cost = replacement costs every 15 years.

to align with the ICMS. Following extensive collaboration with BCIS, the Chartered Institution of Building Services Engineers (CIBSE) and the Building & Engineering Services Association (B&ES) have agreed to adopt the NRM3 expanded cost structure. This means that the NRM3 elemental cost structure is now fully aligned with industry-standard planned preventative maintenance task schedules and the economic reference life expectancy data structure published by CIBSE Guide M and the BCIS cost analysis.

NRM3 RICS new rules of measurement for building maintenance works provide a structured basis for measurement of building maintenance works, encompassing the annualised maintenance and life cycle major repairs and re-placements of constructed assets and building components – which are carried out post construction procurement and throughout the in-use phases of the constructed assets, or built environment. The prime functions of these rules of measurement are to provide consistent rules for the quantification and mea-surement of building maintenance work items – for the purposes of producing order of cost estimates, elemental cost plans and detailed asset specific work programmes, throughout the entire building life cycle.

The secondary functions of these rules of measurement for maintenance works include, amongst others, to provide information for:

1   Input into life cycle cost plans in a structured way so that the same approach is adopted for all life cycle cost plans cash flows and option appraisals. This in turn will facilitate meaningful comparison and more robust data analysis
2   Advising clients on the likely cash flow requirements for the purpose of annual budgeting (and initiating sinking funds) and informing forward maintenance and life cycle renewal plans
3   Informing the implementation of maintenance strategy and procurement stages and cost control of expenditure on maintenance works.

The process of economic evaluation of the whole life cycle costing of all construction, operation and maintenance-related costs during ownership is what is commonly referred to as life cycle costing. It provides a method for quantity surveyors/ cost managers to assist building owners and project teams in selecting the optimum solution for their circumstances and help inform the decision-making process at various stages during the feasibility, design development and procurement and the in-use phases of building or facility. NRM3 does not deal with operation or occupancy costs, or energy/carbon and environmental costs as these are too unpredictable.

According to the RICS, NRM3 can be used for:

•   Initial order of cost estimates during preparation stages
•   Cost plans during the design development and pre-construction stages
•   Asset-specific cost plans during the pre-construction phases.

It also offers guidance about:

•   The procurement and cost control of maintenance works
•   The measurement of other items associated with maintenance works that are not included in work items.

The rules for NRM3 follow the same framework and premise as NRM1: Order of cost estimating and cost planning for capital building works and is divided into five parts with supporting appendices:

•   *Part 1* – This section explains placing order of cost estimating and cost planning in context with the *RIBA Plan of Work* and the *OGC Gateway Process*; defines the purpose, use and structure of the rules; clarifies how maintenance relates to other aspects of life cycle costing; defines the cost categories and definitions that constituents the building maintenance works (renewal and maintain); provides preparation rules for defining the brief and project particular requirements: guidance on the process of cost estimating and cost planning and levels of measurement undertaken depending on the stage in the building life cycle; how to deal with projects comprising multiple buildings or facilities; and explains the symbols, abbreviations and definitions used in the rules

- *Part 2* – This section explains how maintenance costs relate to construction and life cycle costing; defines the scope and parameters for renewal (R) and maintenance (M) cost categories; explains the process of cost estimating and cost planning, and the levels of measurement; and explains the importance of developing a clear and comprehensive client's maintenance requirement, and how the measurement rules are to be applied
- *Part 3* – This section describes the purpose and content of an order of cost estimate; defines its key constituents; explains how to prepare and report an order of cost estimate; and sets out the rules of measurement for preparation of order of cost estimates using the floor areas method, functional unit method and elemental method
- *Part 4* – This section describes the purpose and content of elemental cost planning used for building maintenance works; defines its key constituents; explains the rules for measurement for the preparation and reporting of formal maintenance cost plans for maintenance works
- *Part 5* – This section describes a method of calculating annualised costs for renewal and maintenance works. It also explains the methodology for generating life cycle cost plans and various metrics used for life cycle cost analysis and the benchmarking of maintenance works
- *Part 6* – This section comprises the rules of measurement for the elemental cost planning of maintenance works. It explains the use of tables and describes how to codify elemental renewal and maintenance cost plans. Guidance is also given on how to reallocate costs from elements and sub-elements to work packages where building maintenance works are to be procured through use of discrete work packages and a combination of contract strategies. The tables can also be used as a basis for measuring quantities for the application of life cycle costing.

Key questions to consider during the procurement stages

- Has research been carried out by the design team and/or use of the WRAP Net Waste Tool to identify where on-site waste arises?
- Can construction methods that reduce waste be devised through liaison with the contractor and specialist subcontractors?
- Have specialist contractors been consulted on how to reduce waste in the supply chain?
- Have the project specifications been reviewed to select elements/ components/materials and construction processes that reduce waste?
- Is the design adaptable for a variety of purposes during its life span?

- Can building elements and components be maintained, upgraded or replaced without creating waste?
- Does the design incorporate reusable/recyclable components and materials?
- Are the building elements/components/materials easily disassembled?
- Can a BIM system or building handbook be used to record which and how elements/components/materials have been designed for disassembly?

Public private partnerships (PPPs)

Background and definition

In its widest sense a PPP can be defined as; 'a long term relationship between the public and private sectors that has the purpose of producing public services or infrastructure' (Zitron 2004). One of the most popular PPP models, was the Private Finance Initiative (PFI); a term used to describe the procurement processes by which public sector clients contract for capital intensive services from the private sector. The PFI and its successor PF2 are now discontinued as procurement paths due to adverse criticism particularly from trades unions. PPPs bring public and private sectors together in long term contracts. PPPs encompass, voluntary agreements and understandings, service level agreements, outsourcing. PPPs in the UK have developed and continue to be developed in many forms to suit the needs of particular sectors, for example, education and health, and in some cases, sub-sectors, for example, primary health care. The principal PPP models currently in use in the UK construction sector are shown in Figure 3.5.

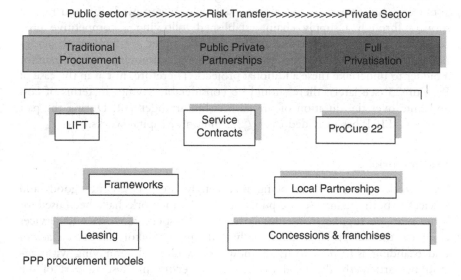

Figure 3.5 PPP procurement models.

In most cases, however, in PPPs arrangements, private sector contractors become the long-term providers of services rather than simply upfront asset builders, combining some or all of the responsibilities for the:

Design
Construction
Finance (which may be a mixture of public and private sources)
Facilities management
Service delivery of a public service facility.

A number of specialist PPP procurement routes have been devised in order to meet the needs of particular public sector agencies, for example;

NHS Local Improvement Finance Trust

Similar to LEPs Local Improvement Finance Trust (LIFT) involves Partnerships UK plc (PUK) and the Department of Health forming a joint venture, Partnerships for Health, to encourage investment in primary care and community-based facilities and services. LIFT has been developed to meet a very specific need in the provision of primary and social healthcare facilities in inner city areas, that's to say GP surgeries, by means of a long-term partnering agreement. In order to participate in the programme, projects must be within areas designated as LIFT by the Department of Health. Although LIFT is at present confined to the health sector other sectors are looking closely at the model for possible adaptation to other public service provision.

LIFT is based on an incremental strategic partnership and is fundamentally about engaging a partner to deliver a stream of accommodation and related services through a supply chain, established following a competitive procurement exercise. Rather like the approach adopted by framework agreements, there should be no need to go through a procurement process again for a bidder to undertake these additional projects. Therefore, just as in the case of ProCure23 (see later), there should be considerable savings in terms of cost and time over the duration of the partnership arrangement. During the past ten years, LIFT has provided over £2.2 billion of capital works.

Frameworks

Framework agreements are being increasingly used to procure goods and services in both the private and public sectors. Frameworks have been used for some years on supplies contracts; however, in respect of works and services contracts, the key problem, particularly in the public sector, has been a lack of understanding as to how to use frameworks, whilst still complying with legislation, particularly the need to include an 'economic test' as part of the process for selection and appointment to the framework. In the private sector, BAA were the first big player to may use of framework agreements and covered everything from quantity surveyors to architects and small works contractors although subsequently BAA change their policy on frameworks with a move towards a move conventional approach to procurement. A framework can be described as *'An agreement between one or more contracting authorities and one or more economic operators, the purpose of which is to establish the terms governing contracts to be awarded during a given period, in particular with regard to price and, where appropriate, the quality envisaged.*

An example of a current successful framework programme is ProCure23.

ProCure23

NHS ProCure23 has been constructed and developed from previous programmes by NHS Estates around four strands to promote better capital procurement by:

- Establishing a partnering programme for the NHS by developing long term framework agreements with the private sector that will deliver better value for money and a better service for patients
- Enabling the NHS to be recognised as a 'Best Client' by providing client training
- Promoting high quality design
- Ensuring that performance is monitored and improved through benchmarking and performance management.

In common with most large public sector providers, the NHS has suffered from the usual problems of schemes being delivered late, over budget and with varied levels of quality combined with little consideration to whole life costs. One of the main challenges to NHS capital procurement is the fragmentation of the NHS client base for specific healthcare schemes, as it comprises several hundreds of health trusts who all have responsibility for the delivery of schemes and each having differing levels of expertise and experience in capital procurement. The solution to these problems was to develop an approach to procurement known as NHS ProCure23 as a radical departure from traditional NHS procurement methods and its' corner stone of the massive capital investment programme in the NHS. The principle underpinning the ProCure23 programme is that of partnering with the private sector construction industry. The frameworks have a four-year life span and since 2016 successive ProCure frameworks have completed 144 major works schemes and 37 small works packages for 112 NHS clients with a capital value of £5016.8 million and it is claimed have delivered over £21 million in savings for NHS trusts.

Public procurement

During the early 1990s, Euromania broke out in the UK construction industry, and 1 January 1993 was to herald the dawn of new opportunity. It was the day the remaining physical, technical and trade barriers were removed across Europe, and from now on Europe and its markets lay at the UK construction industry's feet. Optimism was high within the UK construction industry – after all, it seemed as though a barrier-free Europe with a multi-billion-pound construction-related output, according to the European Construction Industry Federation was the solution to the falling turnover in the UK. Almost every month, conferences were held on the theme of how to exploit construction industry opportunities in Europe. However, in 2020, the UK left the EU and the single market and the UK no longer has automatic access to the European single market. In addition, only a limited range of professional qualifications are automatically recognised by the EU. These are, nurses, midwives, doctors, dentists, pharmacists, architects and veterinary surgeons.

Procurement in the public sector involves Governments, utilities (i.e., entities operating in the water, energy and transport sectors) and local authorities purchasing goods, services and works over a wide range of market sectors, of which construction is a major part. Public procurement is different from private business transactions in several aspects; the procedures and practices are heavily regulated and, whilst private organisations can spend their own budgets more or less as they wish (with the agreement of their shareholders), public authorities receive their budgets from taxpayers and therefore have a responsibility to obtain value for money, traditionally based on lowest economic cost. However, in recent years, the clear blue water between private and public sectors has disappeared rapidly with the widespread adoption of PPPs and the privatisation of what were once publicly owned utilities or entities.

The impact of BREXIT on public procurement in the UK

Procurement law in the UK is generally set out in the following statutory instruments:

- Public Contracts Regulations 2015 (and the Public Contracts (Scotland) Regulations 2015 in Scotland)
- Utilities Contracts Regulations 2016 (and the Utilities Contracts (Scotland) Regulations 2016 in Scotland)
- Concession Contracts Regulations 2016 (and the Concession Contracts (Scotland) Regulations 2016 in Scotland)
- Defence and Security Public Contracts Regulations 2011.

UK public procurement law pre-existed the UK's exit from the EU. The regulations continue in force today, and none have been repealed in full as a consequence of BREXIT.

Each set of regulations has had to be updated because the UK is no longer an EU Member State, including, removal of references to notices being published in the Official Journal of the EU (OJEU). Financial thresholds are still set at EU levels, at least in the immediate term. In addition, contracting authorities/utilities in the UK will still have to advertise their procurement opportunities, follow the procedures that are set out in the regulations (e.g., open, restricted, competitive dialogue and competitive procedure with negotiation, etc.) and observe key principles such as transparency, equal treatment and non-discrimination.

As UK contracting authorities/utilities can no longer advertise their contracting opportunities via the OJEU the UK Government has introduced a new subscription-based Find a Tender Service or FTS and all notices required to be published under the regulations must be published here from the end of the transition period.

What changes are likely in the longer term?

A Government Green Paper was released on 15 December 2020 with several radical proposals for reform including the following:

- Introducing new principles of 'value for money', 'the public good', 'integrity' and 'efficiency'
- Combining the legislation covering public contracts, defence, utilities and concessions procurements into a single uniform set of rules
- Condensing the current spectrum of procurement processes into three simple procedures
- Widening the scope for authorities to exclude bidders based on poor past performance and establishing a debarment 'black-list'
- Reforming the rules on framework agreements, dynamic purchasing systems and qualification systems
- Overhauling the way in which bidders are informed of decisions including removing the requirement for a written debrief at the conclusion of a procurement
- Reforming the legal challenge process, including potentially introducing a tribunal system to deal with specific/discrete issues.

For the time being, and until the proposals in the Green Paper are finalised and implemented, the rules largely remain unchanged.

National procurement policy statement (NPPS)

In June 2021, the UK Government published its National Procurement Policy Statement (NPPS). The four key priorities of the statement are as follows:

- Social value outcomes – delivering value for money does not necessarilly equate to the cheapest and could encompass creating new jobs, businesses, etc.

- Commercial and procurement delivery including their supply chains
- Collaboration – consider working together to achieve economies of scale
- Ensuring capacity and capability including ensuring that entities have staff with the right skills.

For the purposes of legislation, public bodies are divided into three classes:

1   Central Government and related bodies, for example, NHS Trusts
2   Other public bodies, for example, local authorities and universities
3   Public utilities, for example, water, electricity, gas and rail.

The current thresholds effective from 1 January 2022 are shown below and must now align with the World Trade Organizations (WTO) thresholds and be inclusive of VAT; under previous EU rules, procurement thresholds were net of VAT.

- Works £5,336,937.00
- Services £138,760.00
- Utilities £426,995.00

Public procurement procedure

For the present, the following procedure applies;

Award procedures

The quantity surveyor must decide at an early stage which award procedure is to be adopted. The following general criteria apply:

- Minimum number of bidders must be five for the restricted procedure and three for the negotiated and competitive dialogue procedures
- Contract award is made based on lowest price or most economically advantageous tender (MEAT); Note that from April 2014 MEAT may also now include the 'best price-quality ratio' assesses on the basis of qualitative, environmental and /or social aspects linked to the subject matter of the contract
- Contract notices or contract documents must provide the relative weighting given to each criterion used to judge the most economically advantageous tender and where this is not possible, award criteria must be stated in descending order of importance
- MEAT award criteria may now include environmental characteristics, for example, energy savings, disposal costs, provided these are linked to the subject matter of the contract.

The choices are as follows:

Open procedure

Allows all interested parties to submit tenders

Restricted procedure

Initially operates as the open procedure but then the contracting authority only invites certain contractors, based on their standing and technical competence, to submit a tender. Under certain circumstances, for example extreme urgency, this procedure may be accelerated.

Negotiated procedure

In this procedure, contracting authority negotiates directly with the contractor of its choice. Used in cases where it is strictly necessary to cope with unforeseeable circumstances, such as earthquake or flood.

Competitive dialogue

In essence, the competitive dialogue procedure permits a contracting authority to discuss bidders' proposed solutions with them before preparing revised specifications for the project and going out to bidders asking for modified or upgraded solutions. This process can be undertaken repeatedly until the authority is satisfied with the specifications that have been developed. The introduction of this procedure enables

- Dialogue with selected suppliers to identify and define solutions to meet the needs of the procuring body and
- Awards to be made only based on the most economically advantageous basis.

In addition,

- All candidates and tenderers must be treated equally and commercial confidentiality must be maintained unless the candidate agrees that information may be passed onto others
- Dialogue may be conducted in successive stages. Those unable to meet the need or provide value for money, as measured against the published award criteria, may drop out or be dropped, although this must be conveyed to all tenderers at the outset
- Final tenders are invited from those remaining based on the identified solution or solutions
- Clarification of bids can occur pre and post assessment provided this does not distort competition.

Competitive procedure with negotiation

Like Competitive Dialogue, and the existing Negotiated Procedure, it is a competitive process where negotiations are to be carried out with all the bidders still in the procurement. The major change from the current Negotiated Procedure will be that following negotiation on submitted tenders there will be a formal end to the negotiating and bidders will then be invited to submit a revised tender (very much like the tender phase in Competitive Dialogue). Another aspect is that it specifies the extent to which the authority can change its requirements during the process. This approach specifically precludes an authority from making changes to the:

• Description of the procurement
• Part of the technical specifications which define the minimum requirements the award criteria.

However, it acknowledges the right to make changes to other parts of the specification, provided bidders are given sufficient time to make an adequate response.

Other points to note include:

• As with Competitive Dialogue, there will be specific grounds which permit its use, this will include that 'due to specific circumstances related to the nature or the complexity of the works, supplies or services or the risks attaching thereto, the contract cannot be awarded without prior negotiations'
• The minimum number of bidders to be invited is three
• It will be possible to hold the negotiation in stages and reduce the number of bidders at the end of a stage
• The ability to hold an accelerated procedure, currently limited to the Restricted Procedure, will be extended to the new procedure making it possible to use it in cases of urgency
• A bidder's solution or other confidential information is not to be revealed to other bidders without specific consent.

The procedure has much in common with Competitive Dialogue. What distinguishes is that, in Competitive Dialogue, the first phase solutions are developed until the authority considers that it has identified one or more capable of meeting its needs and then seeks to formalise positions in a tender, whereas in the Competitive Procedure with Negotiation, tenders are submitted initially, are then subject to negotiation and then resubmitted to finalise positions.

Innovative partnership: In this partnership solutions are not already available on the market.

Concessions: It sets out a basic framework for the award of works and services concessions in the public and utilities sector, subject to certain exemptions in respect of water (such as the disposal or treatment of sewage). The new regime leaves the choice of the most appropriate procedure for the award of concessions to individual contracting entities, subject to basic procedural guarantees, including:

- Certain minimum time limits for the receipt of applications and tenders
- The selection criteria must relate exclusively to the technical, financial and economic capacity of operators
- The award criteria must be objective and linked to the subject matter of the concession
- Acceptable modifications to concessions contracts during their term, in particular where changes are required as a result of unforeseen circumstances.

There are no multilateral rules governing public procurement. As a result, Governments are able to maintain procurement policies and practices that are trade distortive. That many Governments wish to do so is understandable; Government purchasing is used by many as a means of pursuing important policy objectives that have little to do with economics – social and industrial policy objectives rank high amongst these. The plurilateral Government Procurement Agreement (GPA) partially fills the void. GPA is based on the GATT provisions negotiated during the 1970s and is reviewed and refined at meetings (or rounds) by ministers at regular intervals. Its main objective is to open up international procurement markets by applying the obligations of non-discrimination and transparency to the tendering procedures of Government entities. It has been estimated that market opportunities for public procurement increased tenfold as a result of the GPA. The GPA's approach follows that of the European rules. The Agreement establishes a set of rules governing the procurement activities of member countries and provides for market access opportunities. It contains general provisions prohibiting discrimination as well as detailed award procedures. These are quite similar to those under European regime, covering both works and other services involving, for example, competition, the use of formal tendering and enforcement, although the procedures are generally more flexible than under the European rules. However, GPA does have a number of shortcomings. First, and perhaps most significantly, its disciplines apply only to those World Trade Organisation members that have signed it. The net result is a continuing black hole in multilateral WTO rules that denies access or provides no legal guarantees of access to billions of dollars of market opportunities in both the goods and services sectors. The present parties are the European Union, Aruba, Norway, Canada, Israel, Japan, Liechtenstein, South Korea, the USA, Switzerland and Singapore.

Developments in public procurement

As in the private sector, information technology is the driving force in bringing efficiency and added value to procurement. However, despite the many independent research projects that have been undertaken by the private sector, the findings cannot simply be lifted and incorporated into the public sector due to the numerous UK and European regulations that must be adhered to. Notwithstanding these potential problems and the UK Government has set an ambitious target for the adoption of e-tendering in the public sector. Of all strands of the e-business revolution, it is e-procurement that has been the most broadly adopted, has laid claim to the greatest benefits and accounts for the vast majority of electronic trading.

Electronic tendering

Electronic auctions

The internet is making the use of electronic auctions increasingly more attractive as a means of obtaining bids in both public and private sectors; indeed, it can be one of the most transparent methods of procurement. At present electronic auctions can be used in both open and restricted framework procedures. The system works as follows:

- The framework (i.e., of the selected bidders) is drawn up
- The specification is prepared
- The public entity then establishes the lowest price award criterion, for example with a benchmark price as a starting point for bidding
- Reverse bidding on a price then takes place, with framework organisations agreeing to bid openly against the benchmark price
- Prices/bids are posted up to a stated deadline
- All bidders see the final price.

A criticism of electronic auctions is that it produces the lowest price rather than best value.

Bibliography

Achilles (2021). *Transforming Public Procurement*, Achilles Information Limited.
ARCADIS (2022). *The Business Case for Intelligent Buildings*.
Bassi, R. et al. (2021). *Benefits of Modern Methods of Construction in Housing*, Innovate UK.
BIS RESEARCH PAPER NO. 145 (2013). *Supply Chain Analysis into the Construction Industry*, Department for Business, Innovation and Skills.
CBI (2020). *Fine Margins: Delivering Financial Sustainability in UK Construction*, CBI.
CBI (2021). *New Foundations A Fresh Start for Procurement to Help Construction – And the UK – Build Back Better*, CBI.
Construction Leadership Council (2018). *Procuring for value*, CLC.
Clift, M. and Bourke, K. (1999). *Study on Whole Life Costing*, CRC.
Dixon, T. et al. (2007). *Green Profession? An Audit of Sustainability Tools, Techniques and Information for RICS Members*, RICS London.
Evans, Haryott, Haste and Jones (1998). *Royal Academy of Engineering Paper, "The Long Term Cost of Owning and Using Buildings"*, p. 5.
HM Govt (2020). *The Construction Playbook Government Guidance on Sourcing and Contracting Public Works Projects and Programmes*.

H.M. Government (2008). *Strategy for Sustainable Construction*, Dept for Business Enterprise and Regulatory Reform.

HM Government Construction 2025 (2013). *Industrial Strategy: Government and Industry in Partnership*, HMSO.

HM Treasury (2022). *Appraisal & Evaluation in Central Government (2022)*, HMSO.

Hughes, W., Ancell, D., Gruneberg, S., and Hirst, L. (2004). *Exposing the Myth of the 1:5:200 Ratio Relating Initial Cost, Maintenance and Staffing Costs of Office Buildings*, University of Reading.

IPF Research (2009). *Costing Energy Efficiency Improvements in Existing Commercial Buildings*, Investment Property Forum.

Killip, G. (2020). A reform agenda for UK construction education and practice, *Buildings and Cities*, 1(1), pp. 525–537.

New Models of Construction Procurement (2019). The annual report of the NMCP Working Group to the IPA.

RIBA/NBS (2022). *Construction Contracts and Law Report 2022*, RIBA.

RICS (2007). *Surveying Sustainability; A Short Guide for the Property Profession*, RICS London.

RICS (2009). *Sustainability and the RICS Property Life Cycle*, RICS London.

RICS (2009). Kelly, J, Hunter K, *Life Cycle Costing of Sustainable Design*, RICS Research Report.

RICS (2009). *Renewable Energy*, RICS London.

RICS (2020). *The Futures Report*, RICS.

RICS new rules of measurement (2021). *Order of Cost Estimating and Cost Planning for Building Maintenance Works (NRM3)*: Second Edition, RICS.

Secretary of State for Energy and Climate Change (2009). *The UK Renewable Energy Strategy*, HMSO.

Secretary of State for Energy and Climate Change (2009). *The Low Carbon Transition Plan*, HMSO.

Taylor, T. and Ward, C. (2016). *New Methodology for Generating BREEAM Category Weightings*, BRE.

Williams, B. (2006). *Benchmarking of Construction Costs in the Member States*, BWA

Wrap, Cyril Sweett (2009). *Delivering Higher Recycled Content in Construction Projects*, Waste Resources Action Programme.

4 Digital construction and the quantity surveyor

Siobhan Morrison

Introduction

The construction industry is at an inflection point. Digital technologies are disrupting the industry, providing new opportunities to address the challenges of poor profitability/productivity, project performance, skilled labour shortages and sustainability concerns. The sector, alongside others, is experiencing the Fourth Industrial Revolution (4IR) that will fundamentally alter the way we work; that of digitisation. Whilst digital construction has been evolving over some time and gathering particular traction in recent years, as yet there is no clear absolute definition of 'digital construction' that is universal amongst leading industry bodies, organisations and academia. Indeed, digital construction encapsulates a broad arena of processes, tools, datasets, and technologies and takes many forms. What is agreed, is at the core of digital construction are *digital tools* and *data*, separate entities, yet inextricably linked to one another. Utilising digital tools is, arguably, the most efficient, auditable and effective way of harnessing the vast quantities of data, existing and new, within the built environment sector in order to optimise, streamline and enhance both projects and the practices of the sector. Ultimately, 'construction' concerns the entirety of the lifecycle of a project from inception, early-stage design and planning, through to construction, operation and maintenance, and eventual decommissioning or deconstruction of the physical asset. Beyond, the project and lasting legacy of the physical built asset our sector produces is also that of 'constructing' an attractive and successful sector through effective and safe working practices, sustainable business models and being digitally aware and adept. As a sector, we seek to resolve the inefficiencies and criticisms of our practices and to innovate with the new and exciting tools at our disposal in order to build both an attractive industry and sustainable physical assets for future generations.

Digital tools themselves have permeated throughout society in various forms, from our personal lives and interactions and have been harnessed to greater extents within other business sectors, becoming a near necessity for day-to-day business functions. Different business sectors have harnessed the ability of digital tools and data through successful data analysis to leverage value and inform decision making. The construction sector itself is also looking

DOI: 10.1201/9781003293453-4

towards digitisation in order to maximise and enhance the built environment. This convergence of the built environment sector and digitisation is occurring at the same time as the increased focus upon constructing and living sustainably as a consequence of climate change and governing agendas with consequent industry targets. It is, therefore, likely that within the future of any one of these sectors, and particularly for the built environment, there will be a focus upon these three facets in tandem with one another. It is important to note that it is not a full and wholesale new way of working that will, or needs to, happen immediately, rather, it is the merging of sectors in a series of steps to optimise the construction and legacy of the built environment. Digital processes and data are not a silver bullet. They are not an absolute, yet, they have become a near necessity for survival to harness in order to streamline the processes and rectify inefficiencies that are in existence within our industry today. To ignore them is to be 'left behind'.

This chapter seeks to explore and introduce data and digital tools applicable to the construction sector and the surveyor. It is not an exhaustive and fully detailed exploration; indeed, digital tools and data are occurring at a velocity that are not easily surpassed; however, it seeks to give a working understanding and definitions of the differences within a vast and growing arena, reconciling the digital sector to that of the built environment.

Digitisation in the industry to date

Whilst the generally accepted idea is that construction is an industry slow to change, and perhaps lagging behind other sectors in embracing digitisation, it is important to provide some context and scene setting for the adoption of digitisation that has occurred in the past 100 years or so. Traditionally, construction information was conveyed hard copy on paper via drawings, specifications, contracts and letters. In time, this has evolved in tandem with technological advances. Hand drawn paper drawings and letters became printed documents and still available in hard copy until such time as it evolved beyond paper to 'softcopy' items, for instance, pdf versions of said letters and CAD drawings. More currently, there are project-based collaborative platforms integrating all disciplines, often showing real time progress in 3D visualisations and modelling future phases of the build.

As the technological advances in communications, design and general project management progressed, so too did the working practices of construction and surveyors; however, eventually, there were areas that began to stagnate and lag behind other disciplines. This may be most notable in the significant reliance on Microsoft Excel for surveyors in their daily tasks and also as businesses, which is evidenced in research as well as anecdotally. For instance, Microsoft Excel was launched in 1985 and yet, some 30+ years later, it was reported in research published by the RICS in 2016 that quantity surveyors are overwhelmingly using Excel, favouring it over other discipline specific software and applications. A statistic that may not have changed

Table 4.1 ICT and file evolution

Communications		File Storage and File Sharing
Letter and the written word/ speaking in person		File cabinet (hard copy)
Phonecall – speaking		Floppy disc
Email		CD-ROM
Internal Messaging Systems		USB
External Team Only Messaging Systems		Shared Intranets/Extranets
Video Calls		Cloud Storage

significantly in the intervening years from the published study to present day as the sector continues to struggle and grapple with the resources needed to upskill and successfully implement new technologies. This too appears to be the view of the industry as a whole, whereby KPMG (2016) in their audit report of the industry reported that as a sector, the built environment had not yet fully embraced technology and that *'a mere 8% can be categorised as "cutting-edge visionaries."'*

In many aspects of digital tools and data, it is not yet known what the future may hold; however, industry experts predict that greater digitisation and automation of the processes and operations on site will occur. This will allow for precision construction and safer working environments resulting in a 'smarter' and more sustainable built environment with greater long-term thinking regarding the maintenance and reuse of materials at end of life. To use the progression and evolution of communication through the introduction of Information and Communications Technology (ICT) and that too of file storage and sharing, the changing behaviours in working practises for any sector, not only the built environment, are evident. Contextualising progress through technological interventions with relatable areas of our general, non-discipline specific working practice, can allow us to see the potential and parallels in our more sector specific activities (Table 4.1).

Considering the future of the profession

As the future of the profession is considered, it has been identified that there are skills outside the traditional and normal practise of the construction and surveying professions that will be needed. The RICS have noted in their Future of the Profession report that digital disruption is an opportunity for cross discipline and multidiscipline working to share and learn best practise with the eventual potential for construction and surveying disciplines to evolve and gain new skills that are in keeping with the working practises of a digitised sector. It is recognised that data is being produced at such volume and speed, that to gain meaningful insights from it, it will require data analytics. There is a suggestion that those from the computing sector, particularly including data scientists and analysts, will be key to working with construction disciplines in order to 'make sense' of our data and use the information in the best way possible. Indeed, many of the larger global construction firms have already begun to employ data scientists and analysts in order to leverage value from their in-house data and ultimately gain a competitive edge over their rivals. The RICS report also made reference that the convergence of the built environment sector and the digital sector is not likely to be a conversation of only these two factors. There is the increasing focus upon sustainable development which is concurrent with climate change pledges that is likely to be considered within the future of the digitisation of the industry. These previously thought of as separate factors and conversations converging with one another for the way in which humans use their environment and resources as well as how we choose to live with, and within, physical assets and infrastructure.

Regardless of any particular role held within the built environment sector, for instance, designer, contractor or consultant, all areas are undoubtedly experiencing digitisation of some form and need to be aware and conversant in digitisation. From sole traders, small–medium enterprises and large companies; from quantity surveyors to estimator or commercial surveyor, there will be a change in the day-to-day role. Many businesses either already have or are some way through the phases of their digital strategy. Whether it be the processes we do, the systems we use, the way in which we construct in terms of plant and material, and, equally, *what* we construct, professional disciplines and operatives alike are impacted – not least as clients, stakeholders and end users expect it.

Drivers and challenges of embracing digitisation and data

Digital tools and data although separate are inextricably linked; each digital tool produces data. Every transaction occurring within a digital tool is creating some form of data; for instance, the number of files we store to the number of times we access a particular file, each and every occurrence within the digital world is creating a piece of data. As working practises have become increasingly more digitised, so too has the construction sector itself.

Drivers

In addition to standard business practise becoming more digitised, there are other drivers towards embracing digital tools and processes within the built environment sector which can largely be encapsulated in the following areas; environmental, commercial and the evolution of the professional discipline itself.

Firstly, adoption of digital tools is supported by professional bodies and industry experts alike who are endorsing the use of technology and harnessing of data. Some of the benefits they cite include:

- Innovation
- Changing demographics in the workplace
- Greater efficiency
- Evolution of the role to be contemporary with the demands of today's society and business
- Maintaining standards and quality of service
- Sustainable buildings.

As the role of construction disciplines and the surveyor evolves in order to stay current, there has been somewhat of a digital divide between companies who explore and invest in more digitised methods than those who do not. This is occurring in parallel with another digital divide; that which is occurring between the different generations within the changing demographics of the workplace. There are, of course, exceptions to the rule, yet it is difficult to conceive there would not be a digital divide between the younger generations now entering the sector who have grown up with digital devices as an integral part of their day-to-day lives and those who have integrated them into their lives progressively as the technologies become available. The expectations on use of digital devices for both ends of the spectrum certainly vary and it was recently reported that 'Gen Z' spend an average of eight hours per week assisting the generations of previous years in using technology, whereas the multi-generational workplace, if supported correctly, could be a unique and unrivalled opportunity for mutual growth and knowledge sharing (OSlash, 2022). There is no doubt that digital tools are here to stay for day-to-day working practice; however, we need to advance beyond the generic business tools and innovate with the sector specific tools to create an exciting and attractive sector. Attracting people to the construction industry and showcasing the variety of roles within the sector is not a new issue, the sector has struggled to recruit due to an image problem and reputation for being slow to change. A talent war continues to rage on and the expectations of the younger generations entering the workplace have changed, emboldened by the opportunity that digital connectivity affords and accelerated by the change in working patterns a recent global health pandemic brought. As a sector, we need to embrace the digital tools available to successfully recruit and retain the best talent.

The pressing matter of addressing climate change and its devastating effects has become more prominent within society, from personal endeavours, business strategies to Government legislation, climate pledges and more. With an ever-growing human population, the need for food, shelter and infrastructure increases – particularly in cities, where 75% of the world's production of natural resources is consumed (ARUP and BAM, 2016). The industrial revolution, urbanisation and consumerism has accelerated the use of virgin materials, depleting their stores and resulting in other negative effects such as excessive waste, ecosystem pollution and rising carbon emissions (ARUP, 2016). Globally, the engineering and construction sector is recognised as one of the largest consumers of raw materials and is also recognised to be systemically linear in its processes and approach to delivery (McDonough and Braungart, 2009). Construction and demolition waste can amount to 50% of a nation's total waste and the sector is reputedly responsible for over half of all carbon emissions between its construction activities and the operation of buildings themselves.

Often targeted by climate pledges and Government legislation, the sector is urgently exploring solutions to design out waste and pollution, lower their carbon emissions and extend the lifecycle of its structures, whether existing or new build. Urbanisation, the rapid and exponential growth of the human population, places greater stress on not only the planet's resources but also the current built environment and infrastructure which is, arguably, not always suitable for the needs of a population of this scale. The global community is urgently seeking alternate solutions to the scarcity of the planet's resources due to excessive human consumption and so too does the construction sector. The sector is hoping to extract maximum value from the materials it consumes and by utilising digital tools, it is hoped there is less waste and more effective usage of materials which will have extended use through smarter maintenance, all enabled by digital systems.

In a substantial line of industry reports, contractual evolutions and working practise for the sector, collaborative working is often hailed as enabling more effective and successful delivery of projects, resulting in maximum value for the client and successful margins for the contractor. Digitisation of the industry is seen as an enabler of more collaborative working practises, not least with the introduction of BIM. In using digital platforms, there is an increased transparency and connectivity between the various designers and constructors. Information can be retrieved quickly by the user and is not contingent upon a response from another party.

Challenges

As with all proposed disruptive changes and innovations, there are undoubtedly barriers and challenges that will be encountered.

Resistance to change/culture

One of construction's biggest challenges is its inability to embrace any change quickly. There is an ingrained culture whereby it is slow to adapt and change, whether through personnel resistance or organisational resistance. The reasons are varied; however, typically include time constraints, competing demands and agendas coupled with the low operating and profit margins that make investment into training or digital infrastructure difficult.

Understanding, ability and engagement

The sector as a whole typically does not currently understand or have the analytical skills required and as such, these need to be developed. The understanding of the potential of utilising data has begun to infiltrate our sector at high level as well as perhaps at the grass roots, yet, the majority of the sector is caught in between the spectrum; likened to the pitfalls of 'middle management', the appetite to develop understanding and ability within the digital and data world is halted by the mindset of the middle majority, who are understandably getting the day job done and have little to no time to explore trends beyond the many competing priorities of the day to day.

Investment

It includes investment in IT infrastructure, people and their training, and in the processes and systems. There is a time and money investment requirement from organisations which requires a clear business case and strategy for implementation.

Reliability of data sources

The reliability of data sources within the built environment sector will be a challenge given the fragmented nature of the industry, datasets, stakeholders, complex relationships and commercial sensitivity of information. Each dataset requires to be verified and contextualised to ensure its validity and fitness for purpose.

Interoperability

Ability to exchange and manipulate data freely between disciplines and indeed, even departments within one company, can be complicated due to differing software and IT infrastructure. Consider the challenges inherent and realised evident with BIM implementation whereby interoperability has been somewhat of a barrier for the multidiscipline teams with varying software licenses and requirements – one size does not fit all.

Privacy concerns/ethical considerations

Any given organisation has privacy concerns for their data if it is breached externally or even accessed internally by those who do not have the privacy right to view it. As a sector, in addition to the competitive commercial sensitivity, we are responsible too for the privacy of our network and clients. Client protection and appropriate consideration of safeguarding data is a priority to act in an ethical and professional manner, not least that it is also a legislative responsibility with General Data Protection Regulations (GDPR).

Security and hacking

A significant issue surrounding data is hacking and data breaches, it is regularly in the news and can vary from celebrities to public sector arenas such as the National Health Service (NHS). There is no doubt that hacking systems and data can be a lucrative way for holding a person or organisation to ransom for financial reward and no sector/industry or company type is immune to the threat. Due to the large sums of money involved within construction, and particularly for quantity surveyors affiliated with procurement and payments, construction is a sector enticing to hackers.

System failure

Should there be any form of system failure, there needs to be a secondary plan in place in order to deliver the service required. For context, those who save their documents onto the drive of a particular device, should the device break, lose all of their documents unless they have stored/backed them up in another location that can be accessed from another device. The same can be said for system failures, a backup needs to be in place, such as a secondary server, preferably in a different location.

Natural disasters

Despite the advantages of cloud-based and digital storage, data is ultimately stored somewhere, and that is in data centres. Data centres can be large, vast complexes which are often located in remote areas which are subject to extreme weather conditions, for instance, in deserts. Should anything happen to the data centres, there has to be a secondary backup plan in place.

As we navigate through the beginning steps of digitisation of the sector, it is important to identify and plan an adoption strategy with not only opportunities but also challenges in mind and perhaps more critically, strategies to mitigate them.

Data

Every digital process produces data and data existed prior to digitisation. The inflection point of modern living and data, is the data forms are now of a greater variety and being produced on a volume and speed of never before.

Data creation and form varies, depending on a number of factors:

• Industry/sector
• Organisation
• Purpose
• Requirements
• Legislation

- Standards
- Internal source
- External source.

Examples include:

- Transactions
- Social media exchanges
- Digital processes
- Mobile devices
- Medical notes
- Content
- Sensors
- Downloads.

The myriad of data creation and forms has led to a boom within the field and created occupations such as data scientists as well as businesses operating solely in data processing and analytics. To borrow a premise of knowledge management concerning the relationship of data, information and knowledge, the objective is to inform decision making, thus creating value and leverage to meet the needs of an organisation. The same can be said for the conversation surrounding data and the built environment/construction sector (Figure 4.1).

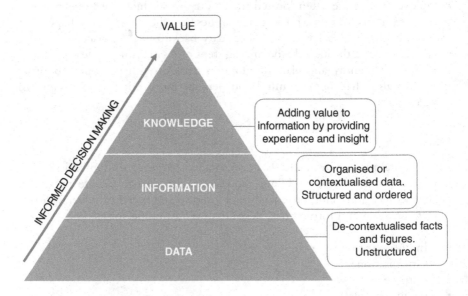

Figure 4.1 Value creation through structured data.

Data representation and terminology

Data representation

It is important to represent data correctly, otherwise it can be misleading. Today, we are saturated with information, particularly visual and audio content. Additionally, this content is becoming shorter and our attention span has decreased as a result whereby we prefer quick and easily interpreted information, often instantly. Consequently, there is a risk that that we can lose context or be misled. Having data in isolation is not necessarily as effective as when it is correlated with other data sets, it is better to incorporate and triangulate with others in order to contextualise and derive the greatest meaning, and anticipated value from it.

For instance, knowing the number of people on a construction site at any given moment is useful, adding in another layer of context could be their trade/discipline with a further layer being the duration of each on the site per day etc. Humans have an unconscious ability to interpret information based on our senses such as visual cues. Representation can be as simple as the size of images in relation to others or utilising a colour coding system – such as a traffic light system, often most associated with risk registers within the construction sector.

Data terminology

Some foundational terminology and definitions for data are provided in Table 4.2 which has been adapted from Scottish Government (2017).

Table 4.2 Data terminology

Terminology	Definition
Administrative Data	Information collected and maintained as part of administration system, e.g., health records, vehicle licensing
Big Data	Large or complex data sets requiring advanced techniques to maintain and analyse
Data Linkage	Joining datasets together
Linked Data	Links to and from other data sets, embedding links, increases exposure
Open Data	Freely available, used and operable
Personal Data	Relating to an individual, individual can be identified
Public Service Information (PSI)	All information collated by public sector tasks, e.g., investment, public realm usage, waste, traffic
Transactional Data	Machine to Machine or Human to Machine interactions., e.g., internet banking, shopping, ATM's, social media interactions

Open data

Open data is freely accessible data that anyone can use, modify, or share and which should not compromise either commercially sensitive or personal information that could be used to identify an individual.

Governments and organisations alike, particularly public sector, recognise the potential of open data to:

- Increase transparency
- Increase civic engagement
- Improve public services
- Boost innovation and economic growth.

Meanwhile, the professional body for open data, The Open Data Institute, state the following benefits to utilising open data:

- Helping Government to make public services more efficient
- Driving innovation and economic growth by revealing opportunities for businesses and start-ups to build new services
- Offering citizens insights into how central and local Government works, improving public trust and boosting political engagement
- Helping Government and communities to keep track of local spending and performance.

The digital, built environment sector and socio-political sectors merge to an extent to empower citizens and optimize services, a sentiment attributed to a smart city. Essentially any recorded information can be a source that is able to become open to all in a useable format, some source examples include addresses, commuter numbers, demographics, carbon footprints, infrastructure investment and any transactions.

Open data sources include Government statistics such as can be found on their website and also the Office for National Statistics (ONS).

Big data

Big data is now a hot topic in businesses and technology has increased greatly over the past few years, the built environment is recognised as having vast quantities of data, in a variety of forms for large, complex projects. The data forms and speed at which it is being produced has been accelerated by digital connectivity and new digital platforms used to foster more streamlined, collaborative and effective working practises throughout the design, planning, construction and operations of a project.

The UK Government define big data as:

> *Big data refers to both large volumes of data with high level of complexity and the analytical methods applied to them which require more advanced techniques and technologies in order to derive meaningful information and insights in real time.*

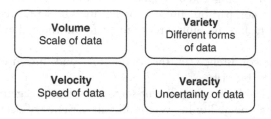

Figure 4.2 The 4 V's of big data.

Essentially, big data is a dataset that cannot be captured, stored, managed or analysed using traditional techniques. New techniques and technology are required to manage and analyse big data effectively in order that patterns and trends can be identified – this is where the value in data analytics lies. There is what is known as the 4 Vs of big data; Volume, Variety, Velocity and Veracity (Figure 4.2).

It is said that with a larger volume of data, and with the speed and variety, it is being created that a more holistic view can be taken to allow for near real-time informed decision making. As the processes and systems solidify in the coming years and as a sector we know what we need from the dataset, the veracity should also improve in tandem.

An important note is how unstructured big data is when no analytics are employed, it is a range of different information in varying formats which must be structured to take any meaning from it. Once analytics are employed, arguably the fifth V of data is created: value.

The construction sector generates vast quantities of data daily whether on a project or organisational level. The sector also benefits from having a wealth of existing data through its documentation procedures already there and able to be harnessed. Whether it is a previous building project and any delays it encountered, and why, or whether certain projects are more prone to delays due to their building type, stakeholders, contractual arrangements, and of course, the costs associated with the construction and maintenance of existing assets.

Some of the data sets currently utilised by construction include the following:

- *Existing data:* plans, planning applications, building warrants, records, archives, estimated and actual costs, delays, risks, case law
- *Environmental data:* Endangered species, historical interest
- Weather reports
- *Geographical:* tunnels, landscapes, locational
- *Health and safety:* statistics, best practice, timelines
- *Community data:* development objections and concerns
- *Traffic reports/communications:* road closures, emergency routes.

Whilst strides are being taken to generate standardised datasets for greater consistency and comparison, unfortunately, an RICS research paper which explored big data and the built environment found the following:

- A lack of consistency in the definitions and measurement of built environment big data
- A low level of built environment sector business engagement
- Lack of interoperability between different varieties of datasets
- The need to better understand the use and supply of built environment big data
- Fragmented ownership of data
- Inconsistency and irregularity in data generation.

It can therefore be concluded that the sector has some way to progress in order to define the needs and capabilities big data may afford the sector. It is, very much, an area in its' infancy for the sector which represents an exciting opportunity for the potential optimisation of the sector through big data.

Digital technologies and concepts

There are a growing number of emerging technologies and digital tools associated with and encapsulated in the umbrella term of 'digital construction'. If not yet disrupting the specific ways of daily working, there is certainly a case for awareness and understanding of the concepts. This section seeks to introduce some of those which are becoming part of the vernacular for the modern quantity surveyor.

Internet of things (IoT)

As the key catalyst to the digital revolution, the internet of things (IoT) concerns the connectivity of 'Things' via devices and software for the purpose of transmitting and exchanging data. In the built environment sector, the 'Things' could range from the built asset in its entirety (a building or piece of infrastructure), building component, material or even to plant and equipment used in the construction phase.

The rise and evolution of Information Communication Technology (ICT), whereby billions of people are connected via devices, coupled with the lower costs of computing equipment and the significant reliance upon them within day-to-day business operations have created this network of connectivity whereby the physical and digital worlds co-exist in parallel. With this, there is an opportunity for the physical world to be enhanced by the information-rich digital IoT.

Artificial intelligence (AI)

AI involves the use of technology iteratively learning the best way in which to optimise its output and environment. By using a dataset and applying set rules and parameters, it will assess whether the result has been achieved, and if not, will learn and adjust in the next attempt. Essentially AI is technology learning through trial and error. Consequently, AI relates to technology problem-solving for decision making which can be termed 'machine learning'. BIM is lauded to be an enabler of AI and has contributed to the focus of AI within the built environment as it provides a dataset and environment which is best suited for AI to utilise and derive meaning from BIM data or optimise it for future needs.

Digital twins

Digital twins comprise a physical built asset and a digital replica of the asset. The digital twin should be updated with real-time data and utilise machine learning to optimise the environment of the physical asset. The use of real-time data is critical as the digital asset should be a complete replica of the physical asset at all points in time, it should not, for instance, be operating on data with a time delay – even that of the previous day. The purpose of the digital twin is to monitor, process, simulate and enable more effective control, use and maintenance of the physical asset. In theory, the building usage, energy requirements, heating and temperature control and even to the timing of opening a door could be monitored and controlled through the digital twin. The digital counterpart utilises machine learning to enable the decision making, learning every second in order to effectively pre-empt and predict the use of the physical asset. It is thought that digital twins could have a positive influence within facilities management.

Augmented reality (AR) and virtual reality (VR)

Often associated with one another, and sometimes incorrectly, used interchangeably are augmented reality (AR) and virtual reality (VR). AR relates to a digital overlay to a physical asset, whereas VR is a digitally generated 3D space which users can interact with. As these technologies become more commonplace, the ability to utilise these technologies to enable clients and stakeholders to visualise their projects has allowed for greater design certainty, thus reducing design changes and variations. Equally, these technologies are being utilised to aid in the construction operations themselves, particularly with benefits associated with health and safety.

Blockchain

Blockchain is a distributed digital ledger which records and tracks transactions amongst a network. A distributed digital ledger is a system for recording these transactions in multiple places in a synchronized way. The transactions may be financial (which may or may not involve cryptocurrency) or contractual in nature. The ability for contractual transactions to be recorded and tracked has also paved the way for 'smart contracts'. Given the nature of the construction sector with multiple contractual relationships, payments and the greater supply chain, it is thought that blockchain could provide a system for tracking all transactions – whether financial or contractual. Its use and prevalence is somewhat contested; however, it did experience a decrease due to the volatility and decreased value of cryptocurrency following the COVID-19 health pandemic. Although not exclusively cryptocurrency, its affiliation with it was sufficient enough to cause concern with the risk averse.

Material passports

Material passports are emerging within the built environment sector and consist of a data rich 'passport' for the components of each building or piece of infrastructure. The 'passport' contains all of the information and data related to the material, for example, its origin, composition and full specification, even including the manner in which it has been constructed, assembled or applied. The usage of material passports is gathering traction as they are seen as an enabler and component of driving circularity within the construction sector. With the information contained within the passport, this should allow for greater opportunity for deconstruction at its end of life. If the material can be recovered and reused for its original purpose, it is maintaining its highest value – a key aspect of the circular economy is circulating resources for their intended purpose for as long a period as possible as discussed in Chapter 2. If the materials are unable to be reused for their intended purpose, they could be recycled, known as downgrading in the circular economy. The information contained in the passport, theoretically, should enable its recovery for reuse, or recycling, based upon its characteristics and construction or assembly.

Robotics

Robotics and robotic automation are being utilised in both on site operations and within off site manufacturing in a bid to enable safer working practices and precision construction. Robotic automated activities relating to cutting of materials reduce waste and extract greater value from the material cost. Activities such as carrying heavy loads, self-driving vehicles, hazardous activities such as welding and site security patrols are becoming commonplace within the sector today.

Sensors

Sensors are being used to greater effect within the built environment. Their use is not solely confined to smart infrastructure and buildings whereby they assist in facilities management and maintenance of assets. They are increasingly being used in the construction phase too, particularly harnessed by commercial organisations.

Some usages of sensors and their advantages are cited as;

- Reducing environmental risks
- Air quality tracking
- Improve urban planning
- Improve traffic management/relieve congestion
- Monitor plant/resource usage
- Improve safety of construction workers
- Track temperature, pressure, acoustics
- Improve maintenance on buildings and infrastructure.

Sensors can be used on plant to monitor their usage and whether they are sitting idle. Companies can use this data to decide whether to hire or purchase a particular plant item and determine which is the most cost efficient for them in the long term. In terms of safety on site for operatives, sensors can be utilised for proximity warnings. Construction workers are using wearable sensors as part of their PPE to improve safety – harmful substances, respiratory, noise and vibrations. Many existing buildings are being retrofitted with this technology to improve their performance and aid in maintenance require-ments. Analysing this data achieved through new technology such as smart sensors can assist in fully informed decision making and increasing value.

Smart cities

The term and concept of a smart city has grown in relevance and popularity for a number of years, based upon, and accelerated, primarily on the increased digitisation and exchange of data at a city level. There is yet to be an absolute agreed definition of the term smart city; however, what is agreed is that at its core as is the merging of physical infrastructure, the digital and data worlds and human participation to deliver a sustainable and inclusive community or city. The definition most used that succinctly describes the concept of a smart city is that of BSI:

> 'Smart city' refers to 'effective integration of physical, digital and human systems in the built environment to deliver a sustainable, prosperous and inclusive future for its citizens'.

(BSI, 2014)

With the convergence of these different stakeholders such as the built en-vironment, local government, the digital sector and end users, it has led to a smart city meaning slightly different things to the different industries, and hence, the varying working definitions. Whilst it may not be a static concept, idea or definition, it is a concept with agreed aims and components. An im-provement in quality of live for citizens should be achieved by optimising the existing, and the new built environment to support daily life. With greater citizen engagement in the management of the city, the city itself should be more prosperous, sustainable and resilient – concepts which are becoming increasingly more pressing to achieve as globally, the human race faces multiple hurdles. The key is harnessing open data and the IoT in order to improve the transportation and logistical operations at city level.

The general principles of a smart city:

- Digital infrastructure
- Internet of Things (IoT)
- 'Smart' technology
- Open reusable data
- Integrated datasets for service delivery
- Enhancing citizens lives
- Placing citizens at the forefront of services received
- Economic, social and environmental benefits
- Sustainable solution to urbanisation.

Asia is predominantly leading the way in terms of new smart cities due to rapidly expanding populations and the need for sustainable living; however, a smart city is not exclusively for newly designed cities and communities, it can also be retrofitting of existing infrastructure. For instance, within the UK and Western Europe, due to land mass reasons and the existing infrastructure, it will be a series of steps to retrofit and implement those areas of the smart city concept where possible. Each city will also vary based upon their own attributes; transport, historical interest, tourism, industry, population and density as well as their geographical location and subsequent environmental considerations.

Some examples of items contained within smart cities are as follows:

- *Shared bicycle schemes* – an integrated system of public bicycles for use with digitised hire, often an app, for the unlocking and docking of bikes after use
- *Electric vehicles* – adequate charging stations must be provided. Again, generally an app which allows for hire, payment, and mapping of which vehicles are available and where they are located
- *Smart lampposts* – Lampposts which are inclusive of electrical vehicle charging points, environmental/traffic monitoring stations, digital street signage, wifi provision and are generally LED
- *Intelligent traffic lights* – smart lights use sensors that assess the volume of traffic on the roads at any given moment and alter the phasing of the red and green lights accordingly. Within peak hours of heavy traffic, the green light could be on for longer to allow more cars to pass through an intersection at one time. Equally, they can be equipped with GPS and used to allow clear paths for emergency service vehicles whilst safely stopping traffic coming from other directions.

Whilst there are many varying and competing agendas as to the development of the smart city concept, it seems likely that the factors of increased digitisation, sustainability and urbanisation may all have played a part in its development and that the built environment sector will play an integral role in the future development.

Smart buildings

On a more granular scale to smart cities, the concept of smart buildings themselves has also grown. They are often called 'Intelligent Buildings' too and utilise digital applications for much of the building operations such as opening windows and doors, or for the mechanical and electrical services such as heating and cooling etc. Similar to and often affiliated with digital twins, they provide optimised functionality of the physical asset, greater efficiencies and reduced emissions. It is not to say that smart buildings, or components of them, is applicable only to new build projects, elements of building intelligence can apply to existing assets too.

Digital measurement

With the dawn of computers, software was developed to aid general business working practises and the construction industry in new, more efficient ways of working. Some of which have specific quantity surveyor application and involvement which can be used throughout the project lifecycle.

One of the areas seen to benefit, or rather can and will benefit most, is in the taking off of quantities – a traditionally time consuming and laborious task.

Maximising efficiency in 'taking off' could lead to allowing greater time to be spent on more value creating activities such as effective cost control through refinement of rates in costing, thorough, unambiguous contract documentation or enhanced focus upon value management and engineering exercises (Figure 4.3).

The industry is overwhelmingly using Excel. This is evidenced in research as well as anecdotally. Research conducted by RICS can be seen demonstrating the use of Excel as the most popular computer software/application used, yet, Excel does not necessarily allow for the most effective digital measurement (unless specific add-ins are used). Digital measurement does require an investment in software and training, however, it has become common amongst most organisations to have some form of digital measurement software. When we speak of digital measurement, it is not merely the process of a digitised take off – that is, a typed version of a take-off, it is the act of measuring dimensions from a soft copy drawing through more specialised applications. The increased link between the digital measurement software increases accuracy and auditability, reduces the margin of error between

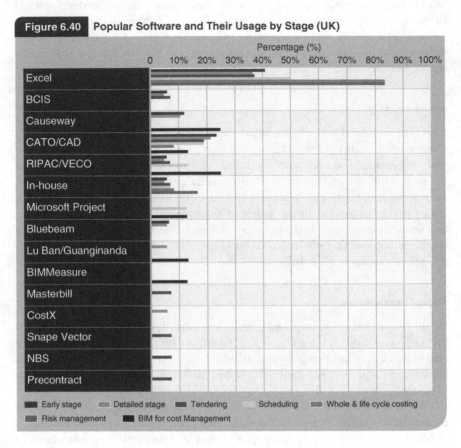

| Figure 6.40 | Popular Software and Their Usage by Stage (UK) |

Figure 4.3 Popular software and their usage by stage (UK) (RICS, 2016).

documents (such as transferral of take-off to bill of quantities/schedule etc.) and allows for more efficient revision if required.

Digital measurement does have its critics, there are some who believe the traditional values are every bit as effective. This is, in part, true. The concept of measurement and the knowledge base required to undertake such activities can be understood via either traditional means or digital means; however, the speed of measurement once digitised is undoubtedly quicker and reduces duplication of taking off dimensions. Traditional measurement conventions and taking off – often on dimension paper – are still dominating much of the surveying education programmes globally; however, we are now seeing the transition and greater inclusion of digital measurement in many institutions. An important note regarding the digital measurement and that of traditional taking off is that the traditional measurement conventions taught do not change simply because a digital process is involved. Traditional conventions referred to are items such as given here:

- A project title
- The person who has undertaken the measurement task for audit purposes
- A date of when the task was undertaken
- A list of information used (including revisions, scales etc.)
- Take off lists
- Dimensions in the order of length, width, height or depth
- Dimensions to two decimal places
- Waste calculations to three decimal places
- Appropriate rounding to demonstrate the quantity to be transferred
- Sign posts – that is, annotated take offs.

The traditional conventions are still employed to ensure a common language for surveyors and the greater industry – i.e., standardisation and consistency.

Digital platforms and databases

Digital platforms and databases have become more and more commonplace within the construction sector and that too is true for the quantity surveyor, whether using a discipline specific platform or engaging in a multi-disciplinary platform.

Perhaps one of the most familiar databases for the quantity surveyor remains the Building Cost Information Services (BCIS). Affiliated with the RICS, the BCIS operates an online subscription service which has undergone a refresh in recent years and provides data on the built environment.

The BCIS database consists of the following sections:

- Analyses
- Indices
- Average prices
- Duration calculator
- Wages
- Dayworks
- Schedule of rates
- Contract percentages
- Tender price studies
- Briefing
- Getting started
- News.

The BCIS relies upon regulated firms and members to submit real project data in order for their analyses, costing information and indices to be updated and valid.

BIM

Multi-discipline and multi-organisational platforms provide a set of tools and a common data environment for collaboration. This can focus on specific functions, such as collaborative design or programming and planning for a particular phase of the asset, say the construction logistics or operations in use phase. The opportunities for built environment software development have resulted in an expanding number of software developers and firms, creating a further parallel competitive marketplace within the sector as they offer their products and services.

One of the greatest platform-based facilitators of information sharing to foster collaborative working practises within the sector, is undoubtedly Building Information Modelling (BIM). Although not an entirely new concept, it has now embedded itself within the industry further from the academic and research world to become a more regular feature and near norm of working practice in the sector. To which degree it is used, how often, and to which effect, is often disputed; however, there is no doubt that it has now established itself as a disruptor of the status quo of printed drawings, pdfs and multiple spreadsheets being circulated by emails.

What is BIM?

BIM is the process of bringing together and sharing information in a digital format among all those involved in a construction project, including architects, engineers, surveyors and constructors. By making information far more accessible and available to the client and end-user to support through-life asset management. BIM is claimed to be a path to greater productivity, more effective risk management, more efficient processes and minimising material usage, ultimately resulting in greater margins and contributing to more sustainable construction. BIM envisages virtual construction of a facility prior to its physical construction, in order to reduce uncertainty, improve safety, clash detection, and simulate and analyse potential impacts. Sub-contractors from every trade can input critical information into the model before beginning construction, with opportunities to pre-fabricate or pre-assemble some systems off site. Waste can be minimised on site and components delivered on a just in time basis rather than stockpiled on site. The use of digital fly-through models is commonplace today; however, with BIM, it's the structured information behind the models that is so important and enables the project to be built digitally before being built on site. On site, it can be used for the purchase of materials and checking compliance of systems and specifications. Once affiliated most closely with a 3D model, this has in time evolved to 4D whereby workflow scheduling was integrated, and then further evolved to integrate cost data, known as 5D. The true adoption of BIM to much beyond 3D or 4D is disputed with very few fully embracing the full potential of 5D BIM models.

Therefore, to put it simply; BIM is a process, facilitated by software which at the higher levels is accessible by all the players in a construction project and is used to enter and store information on a project. The information need not only apply to the construction phase but also to the operation and maintenance of the completed project. In addition, the software can produce virtual models of the complete/proposed project prior to start on site which facilitates:

- Avoidance of design clashes
- Generation of quantities
- The application of alternative solutions including cost estimates
- Sharing of information in real time
- Important information on running and maintenance regimes for the client
- Environmental assessments
- Quicker and easier design revisions
- Consistency of standards
- More accurate scheduling of information
- More accurate tendering processes
- Project planning and resource allocation
- More efficient construction phasing
- Links to facilities management systems.

Why adopt BIM?

Benefits of adopting BIM include:

Feasibility

Allows the best opportunity in early decision-making and greater certainty that project out-comes will be successfully delivered. Collaboration early in the design process reduces the necessity for costly and time-consuming changes and reworks later in the process.

Design stage

Produces simulation models and accurate analysis of design solutions and reduces the potential for clashes resulting in significant de-risk for the client.

Construction/On site

Aids in the planning and construction management of the site logistics (deliveries, routes etc).

Occupancy

Supports facilities management and operation.

End of life

Supports the deconstruction stages and disassembly of assets into materials or components for recovery and reuse.

In order to facilitate the introduction of BIM into the UK construction industry the process has been broken down into levels ranging from 0 to 3 as follows:

BIM Level 0

Level 0 effectively means no collaboration or sharing of information with 2D CAD drafting being utilised, mainly for the production of information with output and distribution being via paper or electronic prints, or a mixture of both. The majority of the industry is already well ahead of this.

BIM Level 1

Working at this level is common within the UK construction industry and generally involves the use of a common data environment, such as a collaboration platform, described earlier in this chapter, with which sharing information can be facilitated and managed. Typically, this level comprises a mixture of 3D CAD and 2D for drafting production information with models not being shared between the project team.

BIM Level 2

This was the level chosen by the UK Government as the minimum target for all work on public-sector work, by 2016. The big difference between level 1 and level 2 is that level 2 involves collaborative working with all parties working and contributing to a shared single model. All parties can access and add to a single BIM model and therefore it is essential that all the parties use a common file format such as COBie (Construction Operations Building Information Exchange). BIM Level 2 was superseded by the UK BIM Framework in 2018, updated in 2021.

BIM Level 3

This level represents full collaboration between all disciplines by means of using a single, shared project model which is held in a centralized repository online. All parties can access a single model and share details and information, including construction phase planning, logistics, costing and asset management.

COBie

Construction Operations Building Information Exchange (COBie) is a non-proprietary data format for the publication of a subset of building information models (BIM) focused on delivering asset data rather that geometric information. It is formally defined as a subset of the Industry Foundation Classes (IFC – the international standard for sharing and exchanging BIM data across different software applications), but can also be conveyed using worksheets or relational databases. It can be used by facilities managers in the operations stage of a project and helps to capture project data such as:

- Equipment lists
- Product data sheets
- Warranties
- Spare parts lists
- Preventive maintenance schedules.

To make COBie as useful as possible, the data is available in several formats and is claimed to be easy to access and manage regardless of size or IT capability.

Challenges for BIM

The following challenges have emerged for early BIM adopters:

Legal issues

These issues centre around:

- Where the design liability lies (if something goes wrong whose fault is it?)
- Ownership – who owns the model and the data in it?
- Copyright and intellectual property rights.

There are differing opinions as to who owns and who should own the intellectual property rights in BIM models. Some consultants accept that these rights are owned by the client while others refuse to hand them over and share BIM data. The main concern being the loss of intellectual property rights. According to *Enabling BIM Through Procurement and Contracts (2016)* King's College Centre of Construction Law and Dispute Resolution, this may be primarily a commercial issue rather than a new legal problem as intellectual property rights should not need additional legal protections by virtue of attaching to BIM models. For example, existing statutory copyright protection already covers graphic and non-graphic design work plus 'computer programmes' and 'preparatory design material for a computer programme'.

Therefore, intellectual property rights and copyright issues would not seem to present a major problem, provided there is a clear understanding as to

- Each team member's ownership or permission in respect of all contributions to models
- Grant of limited, non-exclusive licences to reproduce, distribute, display or otherwise use those contributions

- Equivalent clarity in respect of contractor and sub-contractor contributions; the use of models for facilities management during the operation and maintenance phase.

There is, however, a distinction between intellectual property rights at BIM Level 2, where contributions to BIM models can be traced to their authors, and the position at BIM Level 3, where contributions may become indistinguishable. Concerns have been raised as to the intellectual property implications of BIM Level 3 where contributions cannot be separated and if a contributor cannot prevent or even see amendments made to its work by another contributor. The difference is more a question of insurable liability than intellectual property as it is possible to protect joint authorship if that is how a BIM Level 3 model is to be jointly owned. This situation also raises the question of who pays if information or details in the BIM model turn out to be incorrect.

Who pays for the model?

As previously discussed, the use of BIM is claimed to improve the design, co-ordination and reveals construction problems and therefore helps the design team optimize both product and process and in turn the resultant savings pay for its cost. However, should one party pay for BIM as the savings benefit multiple sources, namely; the design team, main contractors, sub-contractors, suppliers, manufacturers and, of course, the client. The cost of building an integrated model exceeds the normal cost of producing typical construction documentation.

Culture change

At a cultural level, communication, sharing of information and trust has never been the strong point of the UK construction process. For many this factor in the major obstacle to the adoption of BIM by UK plc where to a great extent knowledge is still regarded as power.

What impact will BIM have/BIM had on quantity surveying practice?

Whilst in earlier days, it was suggested that BIM would render quantity surveyors obsolete, as time and BIM adoption has progressed, it can be seen that is not the case. BIM has presented an opportunity for the quantity surveyor to enhance and refine their value in a project team. For instance, the ability to automate time consuming tasks such as taking off and scheduling quantities has allowed quantity surveyors to allocate their time to more value adding tasks such as refining rates/estimating, thus increasing cost certainty or value management and engineering activities to achieve the greatest value for money possible for their client or organisation. The days of spending the larger portion of time measuring and scheduling quantities only to hastily apply rates due to time constraints are eradicated with BIM adoption.

A further opportunity for the quantity surveyor, and a cautionary note for scheduling of quantities from the BIM model, is that of checking the output. The material quantities derived from the model must be verified and checked – currently, a task involving human involvement. The ability of BIM models to automatically generate quantities and cost estimates doesn't lessen the need for an expert to interpret the vast amounts of data produced, or to distil it into a form that clients, contractors and sub-contractors can use to make informed decisions just as in the past they have been used to interpret drawings. Equally, the quality of the data you get out of a BIM model depends on what data you put into it and quantity surveyors are uniquely qualified to input and analyse output of data. Professional insight and judgement will be needed to produce quantities from models and will be for the foreseeable future. There is no doubt that BIM can produce generic quantities more quickly than traditional techniques although other systems such as digitisers are capable of producing quantities quickly and efficiently. However, the key skill for quantity surveyors working with BIM models will be that it is imperative they are able to interpret BIM models, just as in the past they have been used to interpret drawings. BIM models can generate quantities quickly but, what they can't do as effectively just yet is to present the quantities in a format that is of wider use by the industry.

This chapter discussed earlier the convergence of the built environment, digitization and the climate agendas. Again, BIM represents an opportunity for more considered design to reduce waste. Material wastage is said to account for 20–25% of a project's cost, which, when the digital platforms are available to minimize waste from the outset at design stage, it represents an opportunity to utilize digital technologies for more efficient resource use. Following the ability for BIM to aid in the design phases, it can also be utilised within the construction/post contract phases for valuations and payments. This could be managed at either main contract or sub-contract level depending on the project scale, capabilities and appetite to do so. Recording and simulating the build in order to prepare an estimated application for payment can then allow for a comparison to be undertaken at the payment data of forecast versus completed works.

As time and BIM adoption progresses, the full potential is yet to be realised and utilised in day-to-day practice. For the quantity surveyor, whilst many are educated in BIM; be it the protocols, a software itself or the legislation and standards, full integration with costing is not yet the norm.

Conclusion

The growing array of factors that influence our built environment sector and the future digitally aware workforce can be seen as evolutionary rather than revolutionary. Many of the digital tools and datasets discussed are items in existence that have been evolving for some time or are part of the tasks of a surveyor or built environment professional already. What the current rendition of digitisation offers is more sophisticated tools than we had previously, thus affording a chance to streamline and create greater efficiencies to achieve greater value for all involved in the production of the built environment.

As a sector, we have a responsibility to produce lasting quality-built assets for future generations. In doing so, we must minimise our over consumption of the planet's resources and use those we currently have more effectively and for longer. The future generations are not only more climate conscious, but they are also far more digitally adept than we may even be digitally aware, therefore, consideration to the type of built assets, both buildings and infrastructure, that we leave as a legacy, must be future proofed now. If we believe that further merging of the physical and digital worlds is likely in the coming years as the current trajectory suggests, then production of smart buildings and digital twins are what we need to embed now in order that we are equipped.

For an industry attracting new talent and retaining current talent, we need to evolve and embrace newer ways of working. Whilst there are pockets of excellent practice within the sector, digital construction is undoubtedly the present disruptor the wider industry needs to resolve poor productivity, inefficiencies, increase accuracy in our processes, and to provide reliable real time data to clients and project stakeholders. Digital construction not only will have an impact on the quantity surveyor but also affect each and every discipline and each and every organisation in their normal business practice too. If digital tools and data are not embraced now, there is a risk that any given organisation or discipline risks becoming irrelevant and obsolete. What is known, is that the inflection point of the fourth industrial revolution is here, it will fundamentally change the way we work and that it is digital.

References

Arcadis (2022). *The Business Case for Intelligent Buildings*.

ARUP (2016). *The Circular Economy in the Built Environment*. London: ARUP.

ARUP; BAM (2016). *Circular Business Models for the Built Environment*. London: ARUP.

BSI (British Standards Institute) (2014). PAS 180 *The development of a standard on smart city terminology*, BSI.

HM Government Emerging Technologies Big Data Community of Interest (2014). *Emerging Technologies: Big Data*. s.l.: HM Government Horizon Scanning.

King's College Centre of Construction Law and Dispute Resolution *Enabling BIM Through procurement and Contracts* (2016).

KPMG International (2016). *Global Construction Survey 2016: Building a Technology Advantage*. Amstelveen: KPMG International Cooperative.

McDonough, W. and Braungart, M. (2009). *Cradle to Cradle: Remaking the Way We Make Things*. London: Vintage 2009.

OSlash (2022). *Google it First*. [Online] Available at: https://www.oslash.com/google-it-first.

RICS Guidance Note (2015). BIM for Cost Managers: Requirements from the BIM Model.

RICS Futures Programme.

RICS Research (2017) – *Smart Cities, Big Data and the Built Environment: What's Required?*

RICS (2016). *A Comparative Study Of Construction Cost And Commercial Management Services In The UK And China*. [Online] Available at: <https://www.researchgate.net/publication/303567467_A_Comparison_of_UK_and_China_Cost_and_Commercial_Management_Practices/link/5748e27008ae5bf2e63efc9b/download>.

Scottish Government (2017). *A Data Vision for Scotland & Strategic Action Plan. Scottish Government*. [Online] Available at: http://www.gov.scot/Topics/Economy/digital/digitalservices/datamanagement/dmbvfs/dataglossary.

5 Ethics and the quantity surveyor

It is no secret that the construction and infrastructure sector is particularly vulnerable to corruption. Ethics has always been an important topic for the construction industry and particularly so for surveyors who operate in a sector that is generally perceived to have low ethical standards. During the past two decades, society's awareness of the importance of ethical standards appears to have increased due to several high-profile cases across all sectors. Transparency International, a UK-based independent anti-corruption organisation, analyse attitudes to bribery across a range of sectors worldwide. According to Transparency International, corruption and unethical practice erode trust, weaken democracy, hamper economic development and further exacerbate inequality, poverty, social divisions and the environmental crisis. The last time Transparency International included industry sectors in its Bribe Payers Index in 2011, construction ranked first across all types of bribery (petty, grand, and private). There is a wide spectrum of research and models on medical ethics related matters, but comparatively little on business ethics and even less on ethics and construction and surveying practice. Never has there been such a need for individuals and organisations to be seen to be conducting themselves according to ethical principles. For surveying professionals, trust is essential if they are to retain the confidence of the public. However, for some in the construction industry bribery is just part of doing business, *'No one was hurt, the client is happy and the job got finished'*.

A global economic crime survey carried out by PwC found that 49% of all respondents experienced bribery and corruption in engineering and construction compared to 27% overall in other sectors. In addition, 64% of engineering and construction executives saw bribery and corruption as the biggest barrier when operating globally. Respondents also admitted that 29% had been asked to pay a bribe while 38% suspected they had lost contracts to competitors who were willing to pay a bribe. Many other surveys have been conducted in the UK during the past 20 years, notably by the CIOB, with the same depressing conclusions. Interesting 40% of the respondents to a CIOB survey seem to suggest that the practice of cover pricing (i.e. the practice of contractors deliberately submitting a high tender figure or a tender figure linked to numerous caveats to ensure their bid is rejected) is either; *'not very corrupt'* or *'not corrupt at all'* regarding it as the way that the industry operates. One of the major issues from the CIOB survey was a

DOI: 10.1201/9781003293453-5

clear lack of definition of corruption and corrupt practices. The construction industry is one that depends on personal relationships and yet a particular nebulous area is non-cash gifts that range from pens to free holidays. Could it be that construction has become tolerant to corruption?

Corruption is not just bribery; the Global Infrastructure Anti-Corruption Centre (GIACC) includes extortion, fraud, cartels, abuse of power, embezzlement and money laundering in its definition. In a survey carried out in 2020 by YouGov, 97% of middle-market UK construction companies admitted feeling at risk of breaching anti-money laundering and bribery legislation.

The value of global construction output is expected to increase by £6.2 trillion and to reach £13.5 trillion per annum by 2030. It is difficult to precisely determine the value of losses through corruption, but estimates tend to range between 10% and 30%. The experience of the Construction Sector Transparency Initiative (CoST) suggests that a similar amount could be lost through mismanagement and inefficiency. This means that by 2030, unless measures are introduced that effectively improve this situation, close to £4.5 trillion could be lost annually through corruption, mismanagement and inefficiency. So, what is it that makes the construction industry so corrupt?

The potential for bribery and corruption in the construction sector is exacerbated by:

1 Highly competitive tendering processes
2 Low profit margins
3 High levels of debt
4 High levels of insolvencies, (19%, almost one in five, of all UK insolvencies in 2021 were in construction). It is feared that when the pandemic years are analysed, 2022 insolvencies will be even higher.

The above factors exert ever increasing pressure for contractors and sub-contractors to win work. These circumstances mean that companies and individuals are often tempted to engage in bribery to induce procurement decision makers to award work, or to reward them for doing so. The fact that the sector operates via large supply chains means that it is fertile ground for bribery and corruption – there is scope for bribery at each stage of the supply chain process. The Global Infrastructure Anti-Corruption Centre (https://giaccentre.org/) has identified a number of other features that make construction particularly prone to corruption and includes:

Uniqueness: Construction projects are bespoke making comparisons difficult and providing opportunities to inflate costs and conceal bribes.

Complex transaction chains and limited supply chain visibility: The delivery of infrastructure involves many professional disciplines and tradespeople and numerous contractual relationships that make control measures difficult to implement. Construction supply chains can be anything but straightforward. There are often many different suppliers involved and a lot of complex transactions, meaning most businesses don't have end-to-end supply chain visibility. With low supply chain visibility, it can be difficult to keep track of transactions and ensure compliance with anti-corruption and anti-bribery legislation.

Standards of workmanship: Work is concealed or covered up therefore materials and workmanship are often hidden, e.g., steel reinforcement is cast in concrete, masonry is covered with plaster and cables and pipes enclosed in service ducts. Construction projects don't always involve building new structures or giving existing buildings a complete renovation. Work can often be more subtle, for example, if structures are reinforced or facility management tasks are carried out, evidence of this work isn't always visible to the untrained eye. This makes it possible for corruption to occur through unsubstantiated claims of work completed or work completed to an insufficient standard.

Project Costs: Construction projects can cost anything from hundreds to millions of pounds. Every project is unique, meaning every quote given will also be unique. This makes it easier for corrupt individuals or organisations to conceal bribes and inflate costs.

Official bureaucracy: Numerous approvals are required from government in the form of licenses and permits at various stages of the delivery cycle, each one providing an opportunity for bribery.

Consideration of how to combat dishonest or fraudulent activities should be worked into supply chain risk management strategies.

Hiding in plain sight – signs of corruption in construction

Corruption looks different in every organisation and every supply chain, which is why it's not always easy to spot. In some cases, corruption can be hidden so well by those involved that there are almost no identifiable signs. But there will usually be a few subtle indicators that corruption may be an issue within a business. There may be other logical and legitimate explanations for these occurrences. However, by looking out for and investigating these potential indicators of malpractice, you can stay vigilant and reduce the risk of corruption in your organisation. How does you/your organisation measure up? More and more leaders of businesses and other organisations are now waking up to the reality of social responsibility and organisational ethics. Public opinion, unleashed by the social media particularly, is re-shaping expectations and standards. With organisational behaviour, good and bad is more transparent than ever globally and injustice anywhere in the world is becoming more and more visible, and less and less acceptable. As a result, reaction to corporate recklessness, exploitation, dishonesty and negligence it is becoming more and more organised and potent. So how is it possible to recognise unethical practices within an organisation?

- Unnecessary, inappropriate, or poor quality goods or services – this can be a sign of a corrupt relationship with a supplier. Corrupt payments can also be concealed as legitimate or necessary purchases
- Suspicious invoices – false, backdated, or duplicated invoices can be used to conceal illegitimate payments or generate funds to make bribery payments
- Reduced commitment to quality, ethics and compliance – because corruption so frequently involves trying to conceal payments or generate additional funds, the focus of corrupt individuals or practices is usually financial. It's common for commitment to quality, ethics and compliance to take a backstep
- Biased procurement procedures – procurement processes can be vulnerable to corruption and biased decisions can lead to the hiring of suppliers that aren't suitably qualified, trained or skilled. This can deprive other candidates of the right to equal opportunities
- Suppliers refuse to deal with anyone other than their main point of contact. One sign of a corrupt business relationship is if a supplier won't work with anyone other than their usual point of contact. That suppliers prefer dealing with their usual contact is not necessarily unusual, but when suppliers or business contacts don't want to discuss their role with any other members of staff, this could be a sign of corruption
- Individuals reluctant to take annual leave or offer insight into their role – similarly to the above point, individuals within an organisation may be reluctant to use their annual leave for fear of their involvement in corrupt dealings being revealed during the handover to cover staff. They may not want to go into details about certain expenses, projects or supplier relationships
- Sudden increase in work opportunities – marketing efforts, referrals, global events, market trends and seasonal demand can all cause a sudden increase in work opportunities. However, if work suddenly and drastically picks up for no clear reason, this could be an indicator that corruption is at play
- A history of corruption – this isn't to say that leopards can't change their spots, but if an individual or business has a history of being involved in corruption, this could increase the risk of misconduct within your current supply chain
- Lack of written agreements – those involved in corruption will be invested in keeping illegitimate projects, business relationships or expenses a secret. For this reason, there may not be any record or written agreement of contracts that are influenced by corruption.

Many commentators have laid the blame for the attitude of corporates at the feet of business schools who have for years put the financial wellbeing of shareholders at the top of the list of priorities.

The Office of Fair Trading and the Office of National Statistics in one of its largest ever Competition Act investigations found widespread evidence of bid rigging on procurement projects in the UK. The investigation uncovered issues such as so-called 'breakfast-clubs' (a form of illegal cartel where contractors meet

Table 5.1 Types of fraud

Type	Prevalence
Billing fraud	Medium
Bid/contract rigging – market collision	High
Bribery/corruption	Low
Fictitious vendors – false payment applications	Medium
Variation order manipulation	High
Theft or substitution of materials	Medium
False representation	High
Money laundering	Low

up to decide who will win the latest contract). In a 2012 survey of procurement professionals for the AFI, 40% of respondents said that spend on construction is at greatest risk from procurement fraud (Table 5.1).

In some instances, procurement procedures can also inadvertently encourage corrupt behaviour. An investigation by Office of Fair Trading in 2008 uncovered widespread collusion amongst contractors bidding for government contracts and for this reason, particularly in the public sector, there has been a trend to move away from competitive tendering.

The demolition sector's principal trade body is considering making members sign regular anti-collusion declarations to restore trust in the sector following a bid-rigging probe by the Competition and Markets Authority (CMA). In June 2022, the provisional findings of a CMA investigation, begun in 2019, named ten demolition firms, which it said had illegally colluded to rig bids worth more than £150 m over a five and a half-year period. According to the CMA, the firms had agreed to submit bids that were deliberately priced to lose the tender, with several receiving substantial 'compensation payments' worth as much as £500,000. Some firms also produced false invoices to hide their illegal behavior. In all, the CMA said 19 projects were hit by bid-rigging including the Old War Office in London, the redevelopment of Bow Street Magistrates Court, the Metropolitan Police training centre in north London, redevelopment work at Selfridges on Oxford Street, redevelopment work for Oxford University, shopping centres in Reading and Taplow, a new building for the LSE at Lincoln's Inn Fields and the refurbishment of a city office building at 135 Bishopsgate.

In February 2016, the Sweett Group plc was sentenced and ordered to pay £2.25 million as a result of a conviction arising from a Serious Fraud Office investigation into its activities in the United Arab Emirates. The company pleaded guilty in December 2015 to a charge of failing to prevent an act of bribery intended to secure and retain a contract with Al Ain Ahlia Insurance Company (AAAI), contrary to Section 7(1)(b) of the Bribery Act 2010, between December 2012 and December 2015. His Honour Judge Beddoe described the offence as a system failure and said that the offending was patently committed over a period of time. This was the first successful conviction under section 7 of the UK Bribery Act 2010.

ISO 37001 – anti-bribery management system

ISO 37001 is an anti-bribery management system designed to instil an anti-bribery culture within an organisation and implement appropriate controls, which will in turn increase the chance of detecting bribery and reduce its incidence in the first place. ISO 37001, Anti-bribery management systems – requirements with guidance for use – gives the requirements and guidance for establishing, implementing, maintaining and improving an anti-bribery management system. The system can be independent of, or integrated into, an overall management system. It covers bribery in the public, private and not-for-profit sectors, including bribery by and against an organization or its staff, and bribes paid or received through or by a third party. The bribery can take place anywhere, be of any value and can involve financial or non-financial advantages or benefits.

Remote working and ethics

The increased use of remote working since the recent pandemic can pose problems. It is estimated that in 2020, 70% of work forces work remotely at least once a week https://www.toptal.com/. The problem is how does a company create a sense of connection to its purpose and values with new employees who have never met any of their colleagues in person? Are there unique risks that need to be addressed when a company has a large remote workforce or supply chain, following the pandemic of 2020/21? Remote working is becoming an established addition to working practices. One thing is for sure remote working is here to stay. In a recent industry survey, 56% of compliance professionals reported the transition to remote working had gone better than expected, SCCE (2020) *Compliance and the COVID-19 Pandemic, Society of Corporate Compliance and Ethics*. This complements 39% of survey respondents who indicated their risk assessment program did not change as they shifted to remote work. A majority of respondents to the survey also said risk owners had not changed (59%) and the process for remediating risks had remained the same (52%).

Ethics and the law

George McKillop in his paper, *Fraud in Construction – Follow the money isurv Dec 2009*, states 'I wonder, however, how many people fully understand the true diversity of fraud in construction – not only how endemic it is, but it can affect just about any business'. In a highly critical article, McKillop goes onto outline trends in construction fraud giving the following examples:

• The theft and diversion of materials by internal staff
• Quantity surveyors signing off over payments in return for kickbacks

- Construction companies setting up a shell company to invoice a co-operative sub-contractor for non-existent services, which in turn is ultimately billed on to his employer's company for signing off.

The legislative framework defining fraud has been confusing. The principal statutes currently dealing with corruption are the Public Bodies Corrupt Practices Act 1889, the Prevention of Corruption Act 1906 and the Prevention of Corruption Act 1916. This legislation makes bribery a criminal offence whatever the nationality of those involved, if the offer, acceptance, or agreement to accept a bribe takes place within the UK's jurisdiction. The Anti-terrorism, Crime and Security Act 2001 has extended UK jurisdiction to corruption offences committed abroad by UK nationals and incorporated bodies. Commercial bribery is currently covered by the Prevention of Corruption Act 1906 insofar as it relates to bribes accepted by agents. The Bribery and Corruption Act 2010 aimed to *'transform the criminal law on bribery, modernising and simplifying existing legislation to allow prosecutors and the courts to deal with bribery more effectively'*. Additionally, it was hoped that the 2010 Bribery and Corruption Act would also promote and support ethical practice by encouraging businesses to put in place anti-bribery safeguards that ensure all employees are aware of the risks surrounding bribery and that adequate systems exist to manage these. Bribery may include the corruption of a public official as well as commercial bribery, which refers to the corruption of a private individual to gain a commercial or business advantage.

The essential elements of official bribery are:

- Giving or receiving
- A thing of value
- To influence
- An official act.

The thing of value is not limited to cash or money. Such things as lavish gifts and entertainment, payment of travel and lodging expenses, payment of credit card bills, 'loans', promises of future employment, interests in businesses, can be bribes if they were given or received with the intent to influence or be influenced. The act makes it a criminal offence to give, promise or offer a bribe and to request, agree to receive or accept a bribe either at home or abroad. The measures cover bribery of a foreign public official.

The four principal categories of offences in the act are as follows:

1   Offence of bribing another person
2   Offences relating to being bribed
3   Bribery of a Foreign Public Official
4   The new corporate offence: failure to prevent bribery, whereby a commercial organisation (a corporate or a partnership) could be guilty when:

- A bribe has been made by a person performing services for or on behalf of the commercial organisation

- With the intention to obtain or retain business or other business advantage for the commercial organisation.

It is a defence for the organisation to show that there were adequate procedures in place designed to prevent employees or agents committing bribery. The penalties on conviction would be the same as for fraud including, in the most serious cases, a sentence of up to ten years' imprisonment following conviction on indictment.

What has the impact of the 2010 Act been?

The penalties under the act are severe – there is a maximum penalty of 10 years' imprisonment or an unlimited fine for individuals. Corporates face an unlimited fine. However, despite this, the impact of the 2010 act has been disappointing. The case brought against Cyril Sweett is one of a few of successful actions under the 2010 Act. Accountancy and business advisory firm BDO's Fraud Track report, which examines all reported fraud over £50,000 in the UK, found that fraud within construction firms soared by almost £6 million from £2.6 million (2017) to £8.3 million (in 2018). Taking the value of reported fraud in the sector to its highest level since 2010.

Why is ethics important for surveyors?

Professions can only survive if the public retains confidence in them. Conducting professional activities in an ethical manner is at the heart professionalism and the trust that the public has in professions such as the chartered quantity surveyor. One of the principal missions for construction-related institutions like the RICS is to ensure that their members operate to high ethical standards, indeed ethical standards was a top priority on the RICS Agenda for Change (1998). And yet there still appears to be a number of professionals who just don't get it when it comes to ethics, as witnessed by the regular stream of cases that appear before the RICS Disciplinary Panel. In fact, the reported cases are just the tip of the iceberg as many less serious cases brought to the attention of the RICS Professional Conduct Panel are dealt with prior to this stage. For quantity surveyors, transparency and ethical behaviour is particularly important as they deal on a day-to-day basis with, procurement, contractual arrangements and payments and valuations.

The RICS has published a number of guides/documents to help surveyors find their way through the ethical maze:

- Professional ethics guidance note (2000)
- Professional ethics guidance note (2003) – Case studies
- RICS Core Values (2006)
- RICS Rules of Conduct for Members (2007)
- RICS Rules of Conduct for Firms (2007)

- Fraud in Construction – Follow the money (2009)
- Fraud in construction – RICS Guidance (2010)
- RICS Rules of Conduct (2021).

How embarrassing then when the professional body responsible for overseeing ethical practice among its' members, the RICS, was itself found guilty of operating in an unethical manner using some very dubious accounting procedures. In 2019, Sean Tompkins the then chief executive officer (CEO) of the RICS sacked four non-executive directors of the RICS management board for blowing the whistle on the audit by accountancy practice DBO of the institution's 2018 accounts. These events came after *The Sunday Times* reported that findings by BDO, which gave RICS a 'no assurance' rating for the effectiveness of its financial controls in 2018, were not satisfactorily disclosed internally. The ten-page report by BDO had given the RICS the lowest possible 'no assurance' rating for its treasury controls and warned that it was at risk of 'unidentified fraud, misappropriation of funds and misreporting of financial performance'.

The sacked directors had raised concerns that the audit had been suppressed and that sound governance principles were not followed. After several leaks in the press QC Alison Levitt was brought in to investigate the events that led to the dismissal of four directors. Ms Levitt made a number of recommendations as it became apparent that Sean Tompkins had become to regard the RICS as his own personal Fiefdom. The 467-page Levitt Report went onto state:

> *This has been a sad and depressing episode in the life of a great institution. There is a yearning to return RICS to a position of pre-eminence in professional membership organisations. I am confident that with courage and imagination, an independent external governance review will be able to put RICS into the position of moving forward in unity in the public interest.*

As a direct result of the Levitt Report the RICS commenced several initiatives including revising and revisiting the Rules of Conduct. An important point about the RICS Rules of Conduct is that they apply globally, that's to say when working outside the UK the rules still apply and therefore '*When in Rome ...* '. is not a defence for unethical practice or corruption, unlike the CIOB Code of Conduct which states: '*When working in a country other than its own, conduct business in accordance with this Code, so far as it is applicable to the customs and practices of that country*'. The RICS, 2021 Rules of Conduct attempted to simplify the somewhat cumbersome eleven ethical standards into five rules. The Rules of Conduct apply to individuals as well as regulated firms, a departure from the previous rules and contains a list of professional obligations that members have to the RICS. The RICS Rules of Conduct states:

Members and firms must:

1 Be honest, act with integrity and comply with their professional obligations, including obligations to RICS
2 Maintain their professional competence and ensure that services are provided by competent individuals who have the necessary expertise
3 Provide good-quality and diligent service
4 Treat others with respect and encourage diversity and inclusion
5 Act in the public interest, take responsibility for their actions and act to prevent harm and maintain public confidence in the profession.

Additionally, each of the rules is accompanied by a list of 'example behaviours' just in case there is any ambiguity as to how the rules should be interpreted and it would also appear that the rules of conduct have been extended to inclusivity. Although the list above appears to be straightforward, things are never quite that simple in practice when matters such as economic survival and competition are added into the mix. The position is even more complicated when operating in countries outside the UK where business cultures and ideas of ethical behaviours may be very different to those expected by the RICS. Ethical behaviour is that which is socially responsible, for example, obeying the law, telling the truth, showing respect for others and protecting the environment.

Despite resistance from the RICS Governing Council, Lord Michael Bichard was appointed to lead an independent review into the Institution's purpose, governance and strategy. The Bichard Review made firm recommendations on the purpose and governance structure of RICS, and provide advice on values, culture and strategy for an incoming leadership team and Governing Council to take forward. The report, published in June 2022, recommended a five yearly independent review in addition to external performance evaluations for every three years in the light of the fact that the RICS is self-regulating. Lord Bichard expressed the view that he would be surprised in the Government did not consider at some point in the future establishing a separate entity to regulate surveyors.

Established in 2016 the International Ethics Standards Coalition (IESC), on which the RICS is represented, entered the debate with the publication of '*An ethical framework for global property market*' which contained the following list of ethical principles.

Accountability

Practitioners shall take full responsibility for the services they provide, shall recognise and respect client, third party and stakeholder rights and interests and shall give due attention to social and environmental considerations throughout.

Confidentiality

Practitioners shall not disclose any confidential or proprietary information without prior permission unless such disclosure is required by applicable laws or regulations.

Conflict of interest

Practitioners shall make any and all appropriate disclosures in a timely manner before and during the performance of a service. If, after disclosure, a conflict cannot be removed or mitigated, the practitioner shall withdraw from the matter unless the parties affected mutually agree that the practitioner should properly continue. See also; *RICS professional standards and guidance, global Conflicts of interest 1st edition, March 2017* later in this chapter.

Financial responsibility

Practitioners shall be truthful, transparent and trustworthy in all their financial dealings.

Integrity

Practitioners shall act with honesty and fairness and shall base their professional advice on relevant, valid and objective evidence.

Lawfulness

Practitioners shall observe the legal requirements applicable to their discipline for the jurisdictions in which they practice, together with any applicable international laws.

Reflection

Practitioners shall regularly reflect on the standards for their discipline and shall continually evaluate the services they provide to ensure that their practice is consistent with evolving ethical principles and professional standards.

Standard of service

Practitioners shall only provide services for which they are competent and qualified; shall ensure that any employees or associate assisting in the provision of services have the necessary competence to do so and shall provide reliable professional leadership for their colleagues or teams.

Transparency

Practitioners shall be open and accessible, shall not mislead or attempt to mislead; shall not misinform or withhold information as regards products or terms of service; and shall present relevant documentation or other material in plain and intelligible language.

Trust

Practitioners shall uphold their responsibility to promote the reputation of their profession and shall recognise that their practice and conduct bears upon the maintenance of public trust and confidence in the IECS professional organisation and the professions they represent.

Concepts of ethics

Ethical behaviour is developed by people through their physical, emotional and cognitive abilities. People learn ethical behaviour from families, friends, experiences, religious beliefs, educational institutions and media. Business ethics is shaped by societal ethics which, in turn, is a branch of philosophy that covers a whole range of things that have real importance in everyday personal and professional life. Ethics is concerned with the following:

*   Right and wrong
*   Rights and duties
*   Good and bad
*   What goodness itself is
*   The way to live a good life
*   How people use the language of right and wrong.

Ethics tackles some of the fundamentals of life, for example:

*   How should people live
*   What should people do in particular situations.

Therefore, ethics can provide a moral map, a framework that can be used to find a way through difficult professional issues. Business ethics is about the rightness and wrongness of business practices.

Where do ethics come from?

Where do ethics come from; have they been handed down in tablets of stone? Some people do think so and philosophers have several answers to this question when they suggest that ethics originate from;

- God – **Supernaturalism**
- The intuitive moral sense of human beings – **Intuitionism**
- The example of 'good' human beings – **Consequentialism**
- A desire for the best for people in each unique situation – **Situation ethics**.

Ethical dilemmas

Instead of trying to arrive at a standard, all-encompassing rule of what is ethical, it is helpful to illustrate the depth and variety of ethics through suitable examples. An ethical dilemma is a situation in which two or more deeply held values come into conflict. In these situations, the correct ethical choice may be unclear. Perhaps the most common ethical dilemma experienced by quantity surveyors concerns the acceptance of hospitality or gifts from contractors, sub-contractors and clients. In the public sector, the position is somewhat clearer; employees should not accept gifts as they could be construed as a bribe. The common law offence of bribery involves:

> *receiving or offering any undue reward by or to any person whatsoever, in a public office, in order to influence their behaviour in office and incline them to act contrary to known rules of honesty and integrity.*

In addition, Section 1 of the Public Bodies Corrupt Practices Act 1889 makes the bribery of any member, officer, or servant of a public body a criminal offence. In particular, the 1889 Act prohibits the corrupt giving or receiving of:

> *any gift, loan, fee, reward or advantage whatever, as an inducement to, or reward …*

However, in the private sector, the guidelines are not so well defined, for example;

Ethical dilemma

You become aware that the organisation you work for is obtaining work by offering incentives to public officials. Market conditions are highly competitive with little work and many companies are going into receivership. Do you:

- Turn a blind eye?
- Report the matter under the Bribery and corruption Act (2010)?

You are aware that historically whistle blowers great a rough deal.
    Answer is given at the end of this chapter.

Ethics – the business case

Does the financial wellbeing affect attitudes to ethics – is there a business case for ethics? Does ethics add value? Starbucks is one of the world's most ethical companies and makes conscious efforts to be a responsible company and conduct business in an ethical manner. The company is proud of the ethical ways they operate their business through community, sourcing, environment, diversity and wellness. Starbucks has openly displayed its commitment to being socially responsible and they want their customers to know exactly how they are doing. Starbucks led by its CEO Howard Schultz is the largest coffeehouse company in the world with over 33,000 stores in 80 countries. As never before, there are huge organisational advantages from behaving ethically, with humanity, compassion, and with proper consideration for the world beyond the boardroom and the shareholders:

- *Competitive advantage* – Customers are increasingly favouring providers and suppliers who demonstrate responsibility and ethical practices. Failure to do so means lost market share, and shrinking popularity, which reduces revenues, profits, or whatever other results the organisation seeks to achieve
- *Better staff attraction and retention* – The best staff want to work for truly responsible and ethical employers. Failing to be a good employer means good staff leave and reduces the likelihood of attracting good new starters. This pushes up costs and undermines performance and efficiency as has been previously in this book it can cost up to £30,000 to replace a key member of staff. Aside from this, good organisations simply can't function without good people
- *Investment* – Few and fewer investors want to invest in organisations which lack integrity and responsibility, because they don't want the association, and because they know that for all the other reasons here, performance will eventually decline, and who wants to invest in a lost cause?
- *Morale and culture* – Staff who work in a high-integrity, socially responsible, globally considerate organisation are far less prone to stress, attrition and dissatisfaction. Therefore, they are happier and more productive. Happy productive people are a common feature in highly successful organisations. Stressed unhappy staff are less productive, take more time off, need more managing and take no interest in sorting out the organisation's failings when the whole thing implodes
- *Reputation* – It takes years, decades, to build organisational reputation - but only one scandal to destroy it as in the case of Cyril Sweett. Ethical responsible organisations are far less prone to scandals and disasters. And if one does occur, an ethical responsible organisation will automatically know how to deal with it quickly and openly and honestly. People tend to forgive organisations who are genuinely trying to do the right thing.

People do not forgive, and are deeply insulted by, organisations who fail and then fail again by not addressing the problem and the root cause. Years ago, maybe they could hide, but now there's absolutely no hiding place

• **Legal and regulatory reasons** – Soon there'll be no choice anyway - all organisations will have to comply with proper ethical and socially responsible standards and these standards and compliance mechanisms will be global. Welcome to the age of transparency and accountability. So, it makes sense to change before you are forced to

• **Legacy** – Even the most deluded leaders will admit in the cold light of day that they'd prefer to be remembered for doing something good, rather than making a pile of money or building a great big empire. It's human nature to be good. Humankind would not have survived were this not so. The greedy and the deluded have traditionally been able to persist with unethical irresponsible behaviour because there's been nothing much stopping them or reminding them that maybe there is another way. But no longer. A part of the re-shaping of attitudes and expectations is that making a pile of money, and building a great big empire, are becoming stigmatised. What's so great about leaving behind a pile of money or a great big empire if it's been at the cost of others' wellbeing, or the health of the planet? The ethics and responsibility zeitgeist is fundamentally changing the view of what a lifetime legacy should be and can be. And this will change the deeper aspirations of leaders, present and future, who can now see more clearly what a real legacy is.

Ethical decision-making models

The American Accounting Organisation (AAA) incorporated the work of Van Hoose and Paradise (1979), Kitchener (1984), Stadler (1986), Haas and Malouf (1989), Forester-Miller and Rubenstein (1992) and Sileo and Kopala (1993) into a practical, sequential, seven-step, ethical decision-making model. A description and discussion of the steps follows.

1   Identify the Problem.
    Gather as much information as you can that will explain the situation. In doing so, it is important to be as specific and objective as possible. Writing ideas on paper may help you gain clarity. Outline the facts, separating out innuendos, assumptions, hypotheses, or suspicions. There are several questions you can ask yourself: Is it an ethical, legal, professional, or clinical problem? Is it a combination of more than one of these? If a legal question exists, seek legal advice

    Other questions that it may be useful to ask yourself are: Is the issue related to me and what I am or am not doing? Is it related to a client and/or the client's significant others and what they are or are not doing? Is it related to the institution or agency and their policies and procedures? If the problem can be resolved by implementing a policy of an institution or agency, you can look to the agency's guidelines. It is good to remember that dilemmas you face are often complex, so a useful guideline is to examine the problem from several perspectives and avoid searching for a simplistic solution

2 Apply the American Counselling Association (ACA) Code of Ethics

3 After you have clarified the problem, refer to the ACA Code of Ethics (ACA, 2005) to see if the issue is addressed there. If there is an applicable standard or several standards and they are specific and clear, following the course of action indicated should lead to a resolution of the problem. To be able to apply the ethical standards, it is essential that you have read them carefully and that you understand their implications

If the problem is more complex and a resolution does not seem apparent, then you probably have a true ethical dilemma and need to proceed with further steps in the ethical decision-making process

4 Determine the nature and dimensions of the dilemma.

There are several avenues to follow to ensure that you have examined the problem in all its various dimensions:

- Consider the moral principles of autonomy, nonmaleficence, beneficence, justice, and fidelity. Decide which principles apply to the specific situation, and determine which principle takes priority for you in this case. In theory, each principle is of equal value, which means that it is your challenge to determine the priorities when two or more of them are in conflict
- Review the relevant professional literature to ensure that you are using the most current professional thinking in reaching a decision
- Consult with experienced professional colleagues and/or supervisors. As they review with you the information you have gathered, they may see other issues that are relevant or provide a perspective you have not considered. They may also be able to identify aspects of the dilemma that you are not viewing objectively
- Consult your state or national professional associations to see if they can provide help with the dilemma

5 Generate potential courses of action.

Brainstorm as many possible courses of action as possible. Be creative and consider all options and if possible, enlist the assistance of at least one colleague to help you generate options

6 Consider the potential consequences of all options and determine a course of action.

Considering the information you have gathered and the priorities you have set, evaluate each option and assess the potential consequences for all the parties involved. Ponder the implications of each course of action for the client, for others who will be affected and for yourself as a counsellor. Eliminate the options that clearly do not give the desired results or cause even more problematic consequences. Finally, review the remaining options to determine which option or combination of options best fits the situation and addresses the priorities you have identified

7   Evaluate the selected course of action.

Review the selected course of action to see if it presents any new ethical considerations. Stadler (1986) suggests applying three simple tests to the selected course of action to ensure that it is appropriate. In applying the test of justice, assess your own sense of fairness by determining whether you would treat others the same in this situation. For the test of publicity, ask yourself whether you would want your behaviour reported in the press. The test of universality asks you to assess whether you could recommend the same course of action to another counsellor in the same situation

If the course of action you have selected seems to present new ethical issues, then you'll need to go back to the beginning and re-evaluate each step of the process. Perhaps you have chosen the wrong option, or you might have identified the problem incorrectly

If you can answer in the affirmative to each of the questions suggested by Stadler (thus passing the tests of justice, publicity, and universality) and you are satisfied that you have selected an appropriate course of action, then you are ready to move on to implementation

8   Implement the course of action.

Taking the appropriate action in an ethical dilemma is often difficult. The final step involves strengthening your ego to allow you to carry out your plan. After implementing your course of action, it is good practice to follow up on the situation to assess whether your actions had the anticipated effect and consequences.

Source: American Accounting Organisation

Another commonly used ethical decision-making model is;

The American Accounting Association

Again, similar to the previous AAA model this process poses a number of questions.

Step 1: What are the facts of the case?
Step 2: What are the ethical issues in the case?
Step 3: What are the norms, principles, and values related to the case?
Step 4: What are the alternative courses of action?
Step 5: What is the best course of action that is consistent with the norms, principles, and values identified in Step 3?
Step 6: What are the consequences of each possible course of action?
Step 7: What is the decision?

The Laura Nash model

Laura Nash was a senior research fellow at Harvard Business School in 1981 when she drafted a series of 12 questions to help in making difficult ethical decisions as follows:

1 Have you defined the problem accurately?
2 How would you define the problem if you stood on the other side of the fence?
3 How did this situation occur in the first place?
4 To whom and to what do you give your loyalty as a person and as a member of an organisation?
5 What is your intention in making this decision?
6 How does the intention compare with the probable results?
7 Whom could you decision injure?
8 Can you discuss the problem with the affected parties before you decide?
9 Are you confident that your position will be as valid over a long period of time as it seems now?
10 Could you disclose without qualm your decision or action to your boss, the head of your organisation, your colleagues, your family, the person you most admire, or society as a whole?
11 What is the symbolic potential of your action if understood? If misunderstood?
12 Are there circumstances when you would allow exceptions to your stand? What are they?

Of all the decision-making models that have been developed the Nash model would appear to be one of the more respected. Other models include:

Tucker's 5 question model
Mary Guy model (1990)
The Rion model (1990)
The Langenderfer and Rockness Model (1990).

Institute of business ethics

The Institute of Business Ethics (IBE) was established in 1986 to encourage high standards of business behaviour based on ethical values. The IBE defines business ethics as the application of ethical values (such as fairness, honesty, openness, integrity) to business behaviour, for example:

* Are colleagues treated with dignity and respect?
* Are customers treated fairly?
* Are suppliers paid on time?
* Does the business acknowledge its responsibilities to wider society?
* Put simply, business ethics is 'the way business is done around here'.

According to a survey conducted by the IBE (2016) companies regarded the main purpose of having a code of ethical practice as follows:

1	Providing guidance to staff	88%
2	Create a shared and consistent company culture	81%
3	A public commitment to ethical standards	61%
4	Guarding reputation	27%
5	Reducing operational risk	20%
6	Providing guidance to contractors and the supply chain	20%
7	Helping secure long term shareholder value	12%
8	Improving the company's competitive position	7%
9	Decreasing the liability in the case of misconduct	2%

Sustainability and ethics

The question of sustainability is discussed in Chapters 2 & 3.

An international perspective

If ethical dilemmas are problematic on home turf, the problems become magnified and more diverse for construction professionals working outside of the UK. Transparency International is an organisation that seeks to provide reliable quantitative diagnostic tools regarding levels of transparency and corruption, both at global and local levels. Each year they publish a Corruption Perceptions Index that ranks countries and illustrates how countries compare against in other in terms of perceived levels of public sector corruption. Table 5.2 illustrates the results of the 2021 transparency survey. There are 167 counties in the full survey, that can be accessed at https://www.transparency.org/en/cpi/2021

By its very nature, there is little concrete evidence on the levels of bribery and corruption in world market, but companies based in emerging economic giants, such as China, India and Russia, are perceived to routinely engage in bribery when doing business abroad, according to Transparency International's Bribe Payers Index (BPI).

Denmark and New Zealand shared first place in the 2021 BPI with a score of 88 out of a very clean 100, indicating that Danish and New Zealand firms are seen as least likely to bribe abroad. Finland, Singapore, Sweden and Switzerland shared third place on the index, each with a score of 85. At the other end of the spectrum, Somalia and South Sudan ranked last with a score of 12, just below Italy (53), Zambia (33) and Russia (30). Interestingly the BPI also shows public works and construction companies to be the most corruption-prone when dealing with the public sector, and most likely to exert undue influence on the policies, decisions and practices of governments.

The BPI provides evidence that several companies from major exporting countries still use bribery to win business abroad, despite awareness of its

Table 5.2 Transparency international's corruption perception index

Country	ISO3	Region	CPI score 2020	Rank 2020
Denmark	DNK	WE/EU	88	1
New Zealand	NZL	AP	88	1
Finland	FIN	WE/EU	85	3
Singapore	SGP	AP	85	3
Sweden	SWE	WE/EU	85	3
Switzerland	CHE	WE/EU	85	3
Norway	NOR	WE/EU	84	7
Netherlands	NLD	WE/EU	82	8
Germany	DEU	WE/EU	80	9
Luxembourg	LUX	WE/EU	80	9
Australia	AUS	AP	77	11
Canada	CAN	AME	77	11
Hong Kong	HKG	AP	77	11
United Kingdom	GBR	WE/EU	77	11
Austria	AUT	WE/EU	76	15
Belgium	BEL	WE/EU	76	15
Estonia	EST	WE/EU	75	17
Iceland	ISL	WE/EU	75	17
Japan	JPN	AP	74	19
Ireland	IRL	WE/EU	72	20
United Arab Emirates	ARE	MENA	71	21
Uruguay	URY	AME	71	21
France	FRA	WE/EU	69	23
Bhutan	BTN	AP	68	24
Chile	CHL	AME	67	25
United States of America	USA	AME	67	25
Seychelles	SYC	SSA	66	27

damaging impact on corporate reputations and ordinary communities. The Bribe Payers Survey, which serves as the basis for the BPI, also looks at the likelihood of firms in 19 specific sectors to engage in bribery. In the first of two new sectoral rankings, companies in public works contracts and construction; real estate and property development; oil and gas; heavy manufacturing; and mining were seen to bribe officials most frequently. The cleanest sectors, in terms of bribery of public officials, were identified as information technology, fisheries, and banking and finance.

A second sectoral ranking evaluates the likelihood of companies from the 19 sectors to engage in state capture, whereby parties attempt to wield undue influence on government rules, regulations and decision-making through private payments to public officials. Public works contracts and construction; oil and gas; mining; and real estate and property development were seen as the sectors whose companies were most likely to use legal or illegal payments to influence the state. The banking and finance sector is seen to perform considerably worse in terms of state capture than in willingness to bribe public officials, meaning that its companies may exert considerable undue influence on regulators, a significant finding considering the ongoing global financial crisis. The sectors where companies are seen as least likely to exert undue pressure on the public policy process are agriculture, fisheries and light manufacturing however, Transparency International estimated that corruption was costing the construction industry worldwide £3.75 billion per annum.

While most of the world's wealthiest countries already subscribe to a ban on foreign bribery, under the OECD Anti-Bribery Convention, there is little awareness of the convention among the senior business executives interviewed in the Bribe Payers Survey. Governments have a key role to play in ensuring that foreign bribery is stopped at the source – and by making good on commitments to prevent and prosecute such practices.

The FIEC issued a statement in its annual report that the World Bank and the European Union had singled out the construction industry as a sector which, according to their perception, is particularly prone to unethical business practices. As a result, a joint FIEC/EIC Working Group of Ethics was established which published a Statement on Corruption Prevention in the Construction Industry that basically denounced corruption in the construction sector and briefly outlined a code of conduct to promote transparency.

Cross-border construction activity generates opportunities for the construction industry, but it also creates challenges for those involved with the delivery of projects in accordance with local legal requirements and ensuring compliance relevant regulations. In some countries, legal frameworks may be either non-existent, or inadequate as a means of controlling the activities of those involved. The controls imposed through the 1977 Foreign Corrupt Practices Act (FCPA), the 2010 Bribery Act and the Proceed of Crime Act 2002 are flexible enough to operate in relation to construction projects completed anywhere in the world. Any suggestion that they can be ignored, or that compliance may be excused, because of the geographic location, or cultural differences involved will not provide a helpful defence as discussed previously in the case of Cyril Sweett. The regulation of construction projects can be generally said to be less rigorous in high-risk states and there is therefore a temptation for companies and individuals to engage in bribery and corruption as the risk of being discovered is low.

Major construction projects are sometimes carried out in high-risk jurisdictions as emerging markets generates a growing demand for infrastructure and will attracted many new entrants from more established economies the United Kingdom, particularly in the light of Construction 2025, when UK companies have been tasked with expanding into the world construction markets. In some overseas markets local agents are necessary to act as conduits between foreign entities and the local organisations within unfamiliar jurisdictions. Agents may have different views from their employers when it comes to distinguishing between conduct which is regarded as compliant and that which is tainted with bribery, or corruption. Facilitation payments are often required to be paid by contractors in order that imported goods and materials required for a project can clear customs and other government-imposed controls and approvals in relation to plans for design and construction. Given the widespread practice of making facilitation payments in certain countries, it has been questioned whether it is appropriate to criminalise the practice as they help to subsidise the salaries of locally based officials. Consequently, the solution increasingly being adopted by governments of more developed economies is the introduction of their own regulatory controls with extra-territorial reach as with the RICS Code of Conduct. There are several areas where bribery and corruption pose a unique threat to the construction industry. Areas of particular concern to the construction industry include;

Joint ventures

Larger projects are often procured by means of a joint venture, or alliance. The difficulty with these arrangements is that one joint venture partner may have little, if any, control over the acts and omissions of the other partner. Appropriate processes and procedures may have been implemented by one partner and not the other; however, joint venture partners are effectively jointly and severally liable for all their fellow partners.

Nevertheless, for the purposes of the Bribery Act 2010, the existence of the joint venture alone does not automatically mean that one partner is responsible for the other partners. It must also be established that, for example, the joint venture entity was performing services for the joint venture partner and that the joint venture entity paid a bribe for the benefit of that partner. Much will depend on the degree of control which the partners have over each other through contractual and other arrangements. Indeed, where the joint venture itself is conducted through a contractual arrangement, as opposed to a separate legal entity, there is greater likelihood of one partner being held responsible for the acts, or omissions of other partners.

Professional consultants in the UK construction industry have embraced net contribution clauses as such clauses seek to limit a particular consultant's liability to that for which it is directly responsible. But they will not protect against a joint venture partner's failure to prevent bribery by an associated person. Responsibility for such failures cannot be avoided by contract. In the circumstances, the best form of protection is to ensure the joint venture has adequate procedures in place to prevent bribery at the outset.

An ethics charter

One step in raising the awareness of ethical behaviour within an organisation is by drawing up an ethics charter. Some guidelines on implementing this are set out in the following:

1   Get endorsement from the board; corporate values and ethics are matters of governance
2   Find a champion. It is good practice to set up a board level (ethics or corporate responsibility) committee, preferably chaired by a non-executive

director, or to assign responsibility to an existing committee (such as audit or risk). A senior manager will need to be responsible for the development of the policy and code and the implementation of the ethics programme

3   Understand the purpose It is important to clarify the relationship between and understand your organisations approach to corporate responsibility, ethics, compliance and corporate social responsibility strategies

4   Find out what bothers people. Merely endorsing an external standard or copying a code from another organisation will not suffice. It is important to find out what topics employees require guidance on, to be clear what issues are of concern to stakeholders and what issues are material to your business activities, locations and sector

5   Be familiar with external standards and good practice. Find out how other companies in your sector approach ethics and corporate responsibility. Understand what makes an effective policy, code and programme from the point of view of the business, the staff and other stakeholders. How will you embed your code into business practice?

6   Monitoring and assurance. Consider how the success of the policy will be monitored and to whom the business will be accountable regarding its ethical commitments. How will you know it's working? What are the key indicators/measures of an ethical culture for your organisation?

7   Try it out first. The draft code needs piloting – perhaps with a sample of employees drawn from all levels and different locations

8   Review. Plan a process of review that will take account of changing business environments, strategy, stakeholder concerns and social expectations, new standards, and strengths and weakness in your ethical performance.

Another major ethical dilemma that sometimes must be faced is whether to expose malpractice within an organisation for whom you work to the public domain, a practice that has become known as whistleblowing.

Whistleblowing

The unethical behaviour demonstrated by the RICS discussed earlier in this chapter was brought to the public's attention by whistle blowers on the general council. The definition of whistleblowing can be said to be; *'speaking out to the media or the public on malpractice, misconduct, corruption or mismanagement witnessed in an organisation'*. Whistleblowing occurs when a worker raises a concern about danger or illegality that affects others, for example members of the public. Whistleblowing is usually undertaken on the grounds of morality or conscience, or because of a failure of business ethics on the part of the organization being reported. Put at its simplest, whistleblowing occurs when an employee or worker provides certain types of information, usually to the employer or a regulator, which has come to their attention through work. The whistle blower is usually not directly, personally affected by the danger or illegality.

Whistleblowing, as demonstrated by the recent RICS case, is not for the feint hearted as often whistle blowers have been the subject of victimisation, threats, bullying and dismissal. As mentioned earlier there are many case studies on ethical/whistleblowing issues in other sectors, particularly the health sector, but far fewer in construction and related professions. There is, however, one case that came to prominence in 2005 when Production Manager Alan Wainwright, an ex-employee of Haden Young, alleged that the company operated a policy of using a blacklist database. The data base, it was alleged, contained over 500 names of people and companies considered to be disruptive or militant and who should not be hired. In 2005, Wainwright went public on the blacklist issue and shortly afterwards left the company and subsequently failed after 150 job applications to secure a new job, leading him to the conclusion that he himself had been placed on a blacklist. Finally, after 200 applications Alan Wainwright now works as a concert ticket buyer. The consequences for Wainwright were; losing his job, having no income, stress and fear of eviction from his home. In May 2009, Lord Mandleson vowed to introduce legislation to outlaw the compiling and operation of blacklists.

Whist waiting for the promised legislation the key piece of existing whistle blowing legislation is the Public Interest Disclosure Act 1998 (PIDA) which applies to almost all workers and employees who ordinarily work in Great Britain the provisions introduced by the Public Interest Disclosure Act 1998 protect most workers from being subjected to a detriment by their employer. Detriment may take a number of forms, such as denial of promotion, facilities or training opportunities which the employer would otherwise have offered. Under the provisions of the PIDA certain kinds of disclosures qualify for protection ('qualifying disclosures'). Qualifying disclosures are disclosures of information which the worker reasonably believes tend to show one or more of the following matters is either happening now, took place in the past, or is likely to happen in the future:

- A criminal offence
- The breach of a legal obligation
- A miscarriage of justice
- A danger to the health or safety of any individual
- Damage to the environment
- Deliberate covering up of information tending to show any of the above five matters.

It should be noted that in making a disclosure, the worker must have reasonable belief that the information disclosed tends to show one or more of the offences or breaches listed above ('a relevant failure'). The belief need not be correct, it might be discovered subsequently that the worker was in fact wrong, but the worker must show that they held the belief, and that it was a reasonable belief in the circumstances at the time of disclosure.

Employment tribunal statistics show that the total number of people using whistleblowing legislation, which aims to protect workers from victimisation if they have exposed wrongdoing, has increased slowly increasing fears among campaigners that whistle blowers are being deliberately undermined or removed from their workplace, despite repeated promises to protect them.

Whistleblowing procedures

Attitudes towards whistleblowing have evolved considerably during the past 50 years in the early days of the 'organization man' ethos where loyalty to the company was the ruling norm, to the present time when public outrage about corporate misconduct has created a more auspicious climate for whistleblowing. Companies had broad autonomy in employee policies and could fire an employee at will, even for no reason. Employees were expected to be loyal to their organizations at all costs. Among the few exceptions to this rule were unionised employees, who could only be fired for 'just cause', and government employees because the courts upheld their constitutional right to criticize agency policies. In private industry, few real mechanisms for airing grievances existed although, for example, IBM claimed from its earliest days, to have an effective open-door policy that allowed employees to raise any issue. In part because of this lack of protection for whistle blowers, problems were often concealed rather than solved. Probably the clearest examples were in asbestos manufacturing, where the link to lung disease was clearly established as early as 1924 but actively suppressed by company officials for 50 years. In the late 1960s, it was commonplace in the UK to see joiners cutting and drilling asbestos sheet on site, without any form or protection, before fixing it below board flooring as fire protection.

Any whistleblowing policy should aim to:

- Encourage staff to feel confident in raising serious concerns and to question and act upon concerns about practice
- Provide avenues for staff to raise those concerns and receive feedback on any action taken
- Ensure staff receive a response to concerns and that you are aware of how to pursue them if not satisfied
- Reassure staff will be protected from possible reprisals or victimisation if you have a reasonable belief that any disclosure has been made in good faith.

Whistleblowing code of practice

It is important that employers encourage whistleblowing to report wrongdoing and manage risks to the organisation. Employers also need to be well equipped for handling any such concerns raised by workers. It is considered best practice for an employer to:

- Have a whistleblowing policy or appropriate written procedures in place
- Ensure the whistleblowing policy or procedures are easily accessible to all workers
- Raise awareness of the policy or procedures through all available means such as staff engagement, intranet sites, and other marketing communications
- Provide training to all workers on how disclosures should be raised and how they will be acted upon
- Provide training to managers on how to deal with disclosures
- Create an understanding that all staff at all levels of the organisation should demonstrate that they support and encourage whistleblowing
- Confirm that any clauses in settlement agreements do not prevent workers from making disclosures in the public interest
- Ensure the organisation's whistleblowing policy or procedures clearly identify who can be approached by workers that want to raise a disclosure. Organisations should ensure a range of alternative persons who a whistle blower can approach in the event a worker feels unable to approach their manager. If your organisation works with a recognised union, a representative from that union could be an appropriate contact for a worker to approach
- Create an organisational culture where workers feel safe to raise a disclosure in the knowledge that they will not face any detriment from the organisation as a result of speaking up
- Undertake that any detriment towards an individual who raises a disclosure is not acceptable
- Make a commitment that all disclosures raised will be dealt with appropriately, consistently, fairly and professionally
- Undertake to protect the identity of the worker raising a disclosure, unless required by law to reveal it and to offer support throughout with access to mentoring, advice and counselling
- Provide feedback to the worker who raised the disclosure where possible and appropriate subject to other legal requirements. Feedback should include an indication of timings for any actions or next steps.

Source: Dept. for Business Innovation and Skills Guide for Employers – Whistleblowing. March 2015

How not to operate a whistle-blower policy was clearly demonstrated by Barclays in April 2017. Despite the bank having a clear policy towards whistleblowing as follows:

Sometimes the actions of a few may put our reputation at stake. If you believe something is not right – like misconduct, fraud or illegal activity – or if you feel that our standards aren't being met, it is really important that you speak up. Any concerns you may have can be raised in confidence by:

- *Discussing the matter with your manager, or manager's manager*
- *Talking directly to your local Compliance team*
- *Contacting the Whistleblowing team via the Raising Concerns hotline or mailbox.*

> *Concerns raised are taken seriously, treated sensitively, and where appropriate, independently investigated. Where permitted by law, you can raise your concerns with the Whistleblowing team anonymously.*

Barclay's chief executive Jes Staley was severely reprimanded and fined £1.1 million for attempting to discover a whistle blower's identity within the bank following what Mr Staley regarded as *'an unfair personal attack'*. The number of new whistle-blower cases at Barclays dropped by almost a third in 2019, the year after the Jes Staley incident. Jes Staley stood down in 2021 following an investigation into his relationship with Jeffrey Epstein.

Conflict of interest

The RICS publication: *Conflicts of interest RICS professional statement, global 1st edition, March 2017 – effective from 1st January 2018,* is a document that provides members with mandatory requirements that is a rule that a member or firm is expected to adhere to. The professional statement defines 'conflict of interest' as *'a situation in which the duty of an RICS member (working independently or within a non-regulated firm or within a regulated firm) or a regulated firm to act in the interests of a client or other party in a professional assignment conflict with a duty owed to another client or party in relation to the same or a related professional assignment'.*

The statement goes onto suggest that the most important reason for avoiding conflicts of interest is to prevent anything getting in the way of a surveyor's duty to advise and represent each client objectively and independently, without regard to the consequences to another client, any third party, or self-interests and that the clients and in turn the public can have confidence in the system.

Payment on time

Getting paid within a reasonable time frame has always been an issue in construction. Practices such as 'pay when paid' although supposedly outlawed, continues to plague the industry. Payment terms extending to 60,90 or even 120 days are common and a major contributor to the high attrition rates among sub-contractors. To ensure that settlement terms were improved the government introduced a prompt payment code. The code requires firms to uphold best practice for payment standards, including a commitment to pay

95% of all suppliers within 60 days. From 1st September 2019, any supplier who bids for a government contract above £5 million per annum is required to answer questions about their payment practices and performance. Unfortunately, the code appears to have had little effect on payment practice as eight construction companies removed or suspended during the recent past include such household names as Balfour Beatty, Costain, Engie Services, Interserve Construction, Kellogg Brown & Root, Laing O'Rourke and Housebuilder Persimmon.

In January 2021, the code was revised requiring the payment period to small businesses (those with less than 50 employees) to be slashed in half to 30 days, with commitments to be made personally by CEOs or finance director to be effective from 1st July 2021.

Modern slavery

There are different types of modern slavery, but the most common in the construction industry include:

- *Human Trafficking* – A type of organised crime that involves the movement of people using force, fraud, coercion, or deception to exploit them with the offer of well-paid jobs and subsistence. It often involves people being transported internationally, but that's not always the case
- *Forced Labour* – Workers are forced to work against their will and threatened, along with their families, with punishment if they refuse
- *Bonded Labour* – A widespread form of modern slavery where people are forced to work to pay off debts after borrowing money.

The National Crime Agency (NCA) estimates there are tens of thousands of modern slavery and human trafficking victims in the UK. The Global Slavery Index estimates that there are around 136,000 people living in modern slavery in the UK. Slavery is a term used to describe a situation where a person is effectively 'owned' and exploited by others who control where they live and force them to work for little or no pay. Construction in particular; block paving, agriculture, nail bars and car washes were among the top sectors where the Gangmasters operate and Labour Abuse Authority (GLAA) discovered slavery. In 2015, the UK government introduced the Modern Slavery Act which consolidated previous legislation in this area focusing on the prevention and prosecution of modern slavery and the protection of victims. It also sought to invoke transparency in supply chains by requiring all businesses with a global turnover of £36 million or more which supply goods or services in the UK to publish an annual slavery and trafficking statement.

Construction is fertile ground for gang masters to operate due to:

- Convoluted supply chains and subcontracting. Many organisations are unaware of the constituent parts of their supply chains

- The widespread use of self-employment contracts
- Gangmasters operating in the UK are moving workers around internationally to meet shifting demand on building sites
- Business models based on outsourcing (99% of the industry is made up of SMEs)
- Reliance on labour agencies
- A high percentage of migrant workers (one in six of the workforce being born outside of the UK)
- Very low margins (some of the UK's top ten contractors are making less than 2% profit)
- A large proportion of the workforce close to minimum wage
- Lack of labour standards enforcement in the sector.

The Gangmasters and Labour Abuse Authority's Construction Protocol is a joint agreement aimed at establishing collaboration within the construction industry. The Protocol commits signatories to work in partnership to protect vulnerable workers, share information to help stop or prevent exploitation and commit to raising awareness within the supply chain. In 2018, the CIOB published a wide-ranging report entitled: Construction and the Modern Slavery Act: Tackling Exploitation in the UK, in an attempt to both expose the scale and nature of the problem in UK construction as well as alerting the industry to the problem. Evidence of modern slavery in the construction supply chain can be difficult to detect as victims of modern slavery will often try to hide their situation and avoid talking to people out of fear and employers will disguise modern slavery as legitimate employment to avoid prosecution.

Bribery and unethical practice are not confined to the UK. A case was brought by the US Department of Justice, Brazil and Switzerland in December 2016 against Odebrecht and its subsidiary Braskem. The company admitted bribery of $788 million (£553 million) and agreed a record-breaking fine of $3.5 billion. The construction giant paid off politicians, political parties, officials of state-owned enterprises, lawyers, bankers and fixers to secure lucrative contracts in Brazil and abroad. Brazilian politicians were usually paid in cash that was delivered by 'mules', who travelled with shrink-wrapped bricks of banknotes concealed beneath their clothing.

Braskem created a separate department to manage its unethical deals – something prosecutors in Brazil and the United States had never seen before. Between 0.5% and 2% of the corporation's income was moved off-the-books, approximately $600 million a year. The company's bribery department, known by the rather prosaic name of Division of Structured Operations, managed its' own budget. Braskem paid bribes to 415 politicians and 26 political parties in Brazil; however, the web of corruption had tentacles reaching to Africa and across the region forcing the president of Peru to resign and putting the vice-president of Ecuador in prison.

Ethical contracting

Ethical resourcing involves the ethical sourcing of materials and labour and according to ICLEI key elements of ethical sourcing include the following:

- Equal partnership and respect between producers and consumers
- A fair price for socially just and environmentally sound work
- Healthy working conditions
- Fair market access for poverty alleviation and sustainable development
- Stable, transparent and long-term partnership
- Guaranteed minimum wages and prompt payment
- Financial assistance, when needed (pre-production financing)
- Premiums on products used to develop community projects
- Encouraging better environmental practices.

Advice is needed by all types of owners, occupiers, lenders, investors, and public and private bodies as to:

- Their environmental duties and liabilities
- How to determine and quantify liability
- The implications for asset management arising from any actual or potential liabilities
- Who to look to for advice and how advisors should be appointed
- The steps to take to minimise or eliminate liability
- The likelihood of ongoing, new or potential liability.

Non-UK markets

In international terms, the UK construction ranks second in Europe in terms of numbers of people employed and fifth in world rankings (NationMaster.com). Thirteen UK-based companies were amongst the top 50 companies, including Balfour Beatty and the Kier Group. Companies that export tend to be larger, more productive and have greater know how and are more likely to engage in research, development, or wider innovation activity. At the same time, UK construction has a large number of privately owned companies and is more fragmented than its major competitors in Germany or France, driven by a relatively high proportion of self-employment and a relatively high number of small and micro-businesses. In 2021, the value of UK construction industry exports was £3,497 million (ONS). Evidence shows that a relatively small proportion of UK construction contracting firms are exporters comparing to other sectors with about 6% of construction contracting SMEs exporting. Of those contracting SMEs that are non-exporters, about two-thirds said that they did not have a product or a service suitable for exporting and a quarter said that exporting was not part of their business plan. Construction businesses may not be fully aware of potential benefits of exporting and lack the necessary knowledge or management skills to successfully exploit overseas markets. UK construction companies tend to be smaller and collaborate less in comparison to many European countries which may make accessing foreign markets more difficult.

Over the past five years, there have been some dramatic changes in the construction market with businesses across many parts of the world being faced with unprecedented challenges arising from a number of factors. These include rising prices of raw materials, limited availability of funding, corporate failures arising from the inappropriate management of risks, government spending cuts and failing consumer spending coupled with new accounting standards and regulatory requirements. It is anticipated that the global construction market will increase by 4.3% per annum. Demographic change is also expected to have a major impact on such facilities as; healthcare facilities, housing, education and infrastructure. A similar scenario is outlined in the RICS Futures Report 2020.

One of the many factors having to be faced when considering expansion into non-UK markets is the question of differing ethical standards and how these should be navigated. During the past five years, the RICS has increased its' global presence including the appearance of several European and US Universities and Higher Education establishments. However, all eyes are on China, with an economy growing at 3%–4% per annum compared with the average for the Euro Zone of 2% and a year-on-year increase in infrastructure investment such as bridges, factories and power plant.

The multi-cultural team

Quantity surveyors have proved themselves to be adept in a diverse range of skills, often over and above their technical knowledge, with which they serve the needs of their clients. However, when operating in an international environment these skills and requirements are complicated by the added dimension of a whole series of other factors, including perhaps the most influential – cultural diversity. Companies operating at an international level in many sectors have come to realise the importance of a good understanding of cultural issues and the impact that they have on their business operations. In an increasingly global business environment in which the RICS is constantly promoting the surveyor as a global player. It is a fact that the realisation of the importance and influence of cultural diversity is still lacking in many organisations seeking to expand their business outside of the UK.

Today for many quantity surveyors, international work is no longer separated from the mainstream surveying activity; GATT/GPA (Government Procurement Agreement of the World Trade Organisation), etc. are bringing an international dimension to the work of the property professionals. Consultants from the UK are increasingly looking to newer overseas markets such as Europe and regions where they have few traditional historical connections. Consultants must compete with local firms in all aspects of their services, including business etiquette, market knowledge, fees but above all, delivering added value, in order to succeed. With the creation of the single European market in 1993, many UK firms were surprised that European clients, were not at all interested in the novelty value of using UK professionals but continued to award work based on best value for money. As pointed out in Chapter 1, the UK construction industry and its associated professions have ploughed a lonely furrow for the past 150 years or so as far as the status and nature of the professions, procurement and approach to design are concerned, and it could be argued that this baggage makes it even more difficult to align with and/or adapt to overseas markets. Certainly, companies like the French giant Bouygues, with its multidisciplinary *bureau d'études techniques,* have a major advantage in the international markets because of their long-established capability to re-engineer initial designs in-house and present alternative technical offers. Outside Europe, the United States, for example, has seen a decline in the performance of the US construction industry in international markets, a trend that has been attributed in part to its parochial nature in an increasingly global market. Internally, strong trades unions exercise a vice-like grip on the American construction industry.

The construction industry, in common with many other major business sectors, has been dramatically affected by market globalisation. Previous chapters have described the impact of the digital economy on working practices; multinational clients demand global solutions to their building needs, and professional practices as well as contractors are forging international alliances (either temporary or permanent) to meet demand. It is a fast-moving and highly competitive market, where big is beautiful and response time is all important. The demands placed on professional consultants with a global presence are high, particularly in handling unfamiliar local culture, planning regimes and procurement practice, but the reward is greater consistency of workload for consultants and contractors alike. In an increasingly competitive environment, the companies that are operating at an international level in many sectors have come to realise that a good awareness and understanding of cultural issues is essential to their international business performance. Closeness and inter-relationships within the international business community are hard to penetrate without acceptance as an insider, which can only be achieved with cultural and social understanding. In order to maintain market share, quantity surveyors need to tailor their marketing strategies for example, to take account of the different national cultures. Although some differences turn out to be ephemeral, when exploring international markets, there is often a tightrope that must be walked between an exaggerated respect, which can appear insulting, and a crass insensitivity, which is even more damaging. It comes as no surprise that cultural diversity has been identified as the single greatest barrier to business success. It is no coincidence that the global explosion happened just as the e-commerce revolution arrived, with its 365 days a year/24 hours a day culture, allowing round the clock working and creating a market requiring international expertise backed by local knowledge and innovative management systems. Although it could be argued that in a digital age difference are likely to decrease in significance, they are in fact still very important, and remain major barriers to the globalisation of e-commerce. These differences extend also to commercial practice. Even in a digital economy an organisation still needs to discover and analyse a client's values and preferences, and there is still a role for trading intermediaries such as banks, trading companies, international supply chain managers, chambers of commerce, etc. in helping to bridge differences in culture, language and commercial practice. In an era of global markets purists could perhaps say that splitting markets into European and global sectors is a contradiction. However, Europe does have its own unique features, not least its physical link and proximity to the UK, and for some states a single currency.

Exporting construction expertise

The effect of culture on surveyors operating in international markets

As discussed in the opening of this chapter, culture can be a major barrier to international success. Culture must first be defined and then analysed so that it can be managed effectively; thereafter there is the possibility of modelling the variables as an aid to business. A business culture does not change quickly, but the business environment from which it is derived and with which it constantly interacts is sometimes subject to radical and dramatic change. The business culture in a particular country grows partly out of what could be called the current business environment of that country. Yet business culture is a much broader concept, because alongside the impulses that are derived from the present business environment there are historical examples of the business community. For example, as discussed in Chapter 1, the recessions of 1990 and 2008 saw widespread hardship, particularly in the UK construction industry. There have been many forecasts of doom during the early twenty-first century from analysts drawing comparisons between the state of business at this time with that in 1990, when record output, rising prices and full employment were threatening to overheat the economy as well as construction – can there be many quantity surveying practices in the UK that are not looking over their shoulders to see if and when the next recession is coming? Table 5.3 outlines a sample of the responses by 1500 European companies questioned during a study into the effects of culture on business.

So what is culture? Of the many definitions of culture, the one that seems most accurately to sum up this complex topic is *'an historical emergent set of values'*. The cultural differences within the property/construction sectors can be seen to operate at a number of levels, but can be categorised as follows:

1   Business/economic factors – for example, differences in the economic and legal systems, labour markets, professional institutions, etc. of different countries

Table 5.3 The effects of culture on business

China	Cultural differences are as important as an understanding of Asian or indeed other foreign languages.
Far East	One needs to know etiquette/hierarchical structure/manner of conduct in meetings.
Germany	Rigid approach to most operational procedures.
Middle East	Totally different culture – time, motivation, responsibility.
Russia	Inability to believe terms and conditions as stated really are what they are stated to be.
SE Asia	Strict etiquette of business in S. Korea and China can be a major problem if not understood.
France	Misunderstandings occurred through misinterpretation of cultural differences.

2   Anthropological factors, as explored by Hofstede (1984). The Hofstede IBM study involved 116 000 employees in 40 different countries and is widely accepted as being the benchmark study in this field.

Of these two groups of factors, the first can be regarded as mechanistic in nature, and the learning curve for most organisations can be comparatively steep. For example, the practice of quantity surveyors, or the equivalent, in many countries of paying a contractor money in advance of any works on site may seem anathema but for some it is usual practice. It is the second category of cultural factors, the anthropological factors, that is more problematic. This is particularly so for small and medium enterprises, as larger organisations have sufficient experience (albeit via a local subsidiary) to navigate a path through the cultural maze.

Perhaps one of the most famous pieces of research on the effects of culture was carried out by Gert Hofstede for IBM. Hofstede identified four key value dimensions on which national culture differed (Figure 5.1), a fifth being identified and added by Bond in 1988 (Hofstede and Bond, 1988). These value dimensions were power difference, uncertainty avoidance, individualism/collectivism, and masculinity/femininity, plus the added long-/short-termism. Although neatly

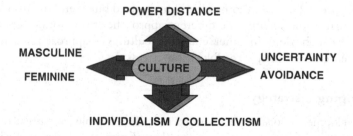

Figure 5.1 Hofstede's four key value dimensions.

categorised and explained, these values do of course in practice interweave and interact to varying degrees.

- **Power distance** indicates the extent to which a society accepts the unequal distribution of power in institutions and organisations, as characterised by organisations with high levels of hierarchy, supervisory control and centralised decision-making. For example, managers in Latin countries expect their position within the organisation to be revered and respected. For French managers, the most important function is control, which is derived from hierarchy
- **Uncertainty avoidance** refers to a society's ability to cope with unpredictability. Managers avoid taking risks and tend to have more of a role in planning and co-ordination. There is a tendency towards a greater quantity of written procedures and codes of conduct. In Germany, managers tend to be specialists and stay longer in one job and feel uncomfortable with any divergence between written procedures – for example the specification for concrete work and the works on site. They expect instructions to be carried out to the letter
- **Individualism/collectivism** reflects the extent to which the members of a society prefer to take care of themselves and their immediate families as opposed to being dependent on groups or other collectives. In these societies, decisions would be taken by groups rather than individuals, and the role of the manager is as a facilitator of the team (e.g., Asian countries). In Japan tasks are assigned to groups rather than individuals, creating stronger links between individuals and the company
- **Masculinity/femininity** refers to the bias towards an assertive, competitive, materialistic society (masculine) or the feminine values of nurturing and relationships. Masculine cultures are characterised by a management style that reflects the importance of producing profits, whereas in a feminine culture the role of the manager is to safeguard the wellbeing of the work force. To the American manager, a low head count is an essential part of business success and high profit; anyone thought to be surplus to requirements will be told to clear his/her desk and leave the company.

As a starting point for an organisation considering looking outside the UK for work, Figure 5.2 may be a somewhat light-hearted but useful discussion aid to help recognise and identify the different approaches to be found towards organisational behaviour in other countries/cultures – approaches that if not recognised can be a major roadblock to success.

Developing a strategy

The development process, when carried out internationally, is particularly complex to manage due to the weaving together of various cultures, including language (both generic and technical), professional standards and construction codes,

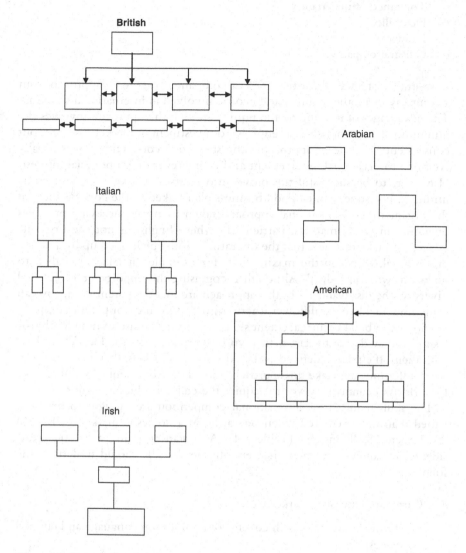

Figure 5.2 Organisational chart.

Source: Adapted from International Management. Reed Business Publishing.

design approaches and technology, codes of conduct, and ethical standards. Technical competency and cultural integration must be taken as read. The competencies necessary to achieve cultural fluency can therefore be said to be:

- Interpersonal skills
- Linguistic ability
- Motivation to work abroad

- Tolerance of uncertainty
- Flexibility
- Respect
- Cultural empathy.

Case studies of SMEs show that 60% of companies react to an approach from a company in another country to become involved in international working. The advantages of reacting to an enquiry are that this approach involves the minimum amount of risk and requires no investment in market research, but consequentially it never approaches the status of a core activity, it is usually confined to occasional involvement and is only ever of a superficial interest. However, to be successful the move into overseas markets requires commitment, investment and a good business plan linked to the core business of the organisation. A traditional approach taken by many surveying practices operating in world markets, particularly where English is not the first language, is to take the view that the operation should be headed up by a native professional, based on the maxim that, for example, 'it takes an Italian to negotiate with an Italian'. Although recognising the importance of cultural diversity, the disadvantages of this approach are that the parent company can sometimes feel like a wallflower, there is no opportunity for parent company employees to build up management skills, and in the course of time local professionals may decide to start their own business and take the local client base with them. If culture is defined as shared values and beliefs, then no wonder so many UK companies take this approach. How long, for example, would it take for a British quantity surveyor to acquire the cultural values of Spain?

The skills required for international competition are thought to be more varied than those required when operating in a domestic market. (UKCES) 2012 Sector Skills Insights (Table 5.4). As a starting point, a practice considering expanding into new markets outside the UK should undertake the following:

1   Carry out extensive market research:

- Ensure market research covers communication (language and cultural issues)

Table 5.4 Reasons given by SMEs for not exporting

Reasons for not exporting	*% of contracting SMEs*
Do not have a service suitable for exporting	66%
Not part of business plan	26%
Have sufficient UK business	11%
Too costly	3%
Lack of time to pursue export opportunities	2%

Source: BIS Small business survey.

- Make frequent visits to the market; it shows commitment rather than trying to pick up the occasional piece of work
- Use written language to explain issues, since verbal skills may be less apparent
- Use exhibitions to obtain local market intelligence and feedback

2 Ensure documentation is culturally adapted and not literally translated:

- Brochures should be fully translated into local language on advice from local contacts
- Publish new catalogues in the local culture
- Set up the web site in the local language, with the web manager able to respond to any leads – after all, if a prospective client is expecting a fast response, waiting for a translator to arrive is not the way to provide it
- Adapt the titles of the services offered to match local perceptions
- Emphasise added value services

3 Depending on the country or countries being targeted, operate as, for example, a European or Asian company, rather than a British company with a multilingual approach – think global, act local:

- Arrange a comprehensive, multi-level programme of visits to the country
- Set up a local subsidiary company or local office or, failing that, set up a foreign desk inside the head office operating as if it is in the foreign country (keeping foreign hours, speaking foreign language, etc.)
- Change the culture of the whole company at all levels from British to European, Asian, etc. as relevant
- Recruit local agents that have been educated in the UK, so they have a good understanding of UK culture too

4 Implement a whole company development strategy:

- Language strategy should be an integral part of a company's overall strategy as a learning organisation
- Identify the few individuals who can learn languages quickly and build on this
- Create in-house language provision
- Set up short-term student placements in the UK for foreign students, via a sponsored scheme
- Target markets whose specialist language ability gives a competitive edge, e.g., China

5 Subcontract the whole export process to a specialist company:

- Hire a company to provide an export package of contacts, liaison, translation, language training, etc

6   Pool resources with other companies:

- Share language expertise and expenses with other companies

7   In joint ventures, collaboration can be based on:

- Equity/operating joint ventures, in which a new entity is created to carry out a specific activity. Seen as a long-term commitment, the new entity has separate legal standing
- Contractual ventures, in which no separate entity is created and instead firms co-operate and share the risk and rewards in clearly specified and predetermined ways. On the face of it, this form of joint venture appears to be more formal

8   Management contracts:

- The transfer of managerial skills and expertise in the operation of a business in return for renumeration.

Conclusion

With the advent of electronic communications, the possibilities that exist for quantity surveyors to operate on a global level have never been greater or easier to access. However, despite what some multinational organisations would have us believe, the world is not a bland homogeneous mass and organisations still need to pay attention to the basics of how to conduct interpersonal relationships if they are to succeed.

Ethical dilemma solution

- Report the matter under the Bribery and Corruption Act (2010).

RICS Members' confidential helpline is available to members on a 9–5 daily basis on +44 20 7695 1670 and one of the range of matters that members can get help and advice on is ethical dilemmas.

Bibliography

Building, 9 July 2022, *Trade body could make demolition firms sign anti-collusion declaration*.

Bardouil, S. (2001). Surveying takes on Europe. *Chartered Surveyor Monthly*, Jul/Aug, pp. 20–22.

Brooke, M. Z. (1996). *International Management*, 3rd edn. Stanley Thornes.

Cartlidge, D. (1997). It's time to tackle cheating in EU public procurement. *Chartered Surveyor Monthly*, Nov/Dec, pp. 44–45.

Cartlidge, D. and Gray, C. (1996). *Cross Border Trading for Public Sector Building Work within the EU*, European Procurement Group, Robert Gordon University.

CIOB (2006). *Corruption in the Construction Industry – A Survey*, The Chartered Institute of Building.

CIOB (2018). *Construction and the Modern Slavery Act Tackling: Exploitation in the UK*, The Chartered Institute of Building.

Dept. for Business Innovation and Skills Guide for Employers (2015). *Whistleblowing*.

Ethics & Compliance Initiative (2021). *Ethics & Compliance Considerations in a Remote Working Environment*.

Evans, R. (2009). *Alan Wainwright: The lonely life of the construction industry whistleblower*, The Guardian May 15th.

Fisher, C. and Lovell, A. (2009). *Business Ethics and Values Third Edition*, Prentice Hall.

Gangmasters and Labour Abuse Authority (2020). *Gangmasters and Labour Abuse Authority's Construction Protocol*.

Hagen, S. (ed.) (1997). *Successful Cross-Cultural Communication Strategies in European Business*, Elucidate.

Hagen, S. (1998). *Business Communication Across Borders: A Study of the Language Use and Practice in European Companies*, The Centre for Information on Language Teaching and Research.

Hall, E. T. (1990). *Understanding Cultural Differences*, Intercultural Press.

Hall, M. A. and Jaggar, D. M. (1997). Should construction enterprises work internationally, take account of differences in culture? Proceedings of the Thirteenth Annual ARCOM 97 Conference.

Hofstede, G. (1984). *Culture Consequences: International Differences in Work-Related Values*, Sage Publications.

Hofstede, G. and Bond, M. H. (1988). *The Confuscius Connection: From Cultural Roots to Economic Growth*, Organizational Dynamics.

International Ethics Standards Coalition (IESC) (2016). *An Ethical Framework for Global Property Market*, King's College Cambridge, 15–17 September.

Matthews, D. (2010). Cover pricing 'as common now as it was two years ago', *Building*, 11th June, p. 15.

Moore, G. and Robson, A. (2002). *The UK Supermarket Industry; and Analysis of Corporate and Social Responsibility*.

Nash, L. (1981). Ethics without the Sermon, *Harvard Business Review*, 59 Nov/Dec 1981.

Plimmer, F. (2009). *Professional Ethics – The European Code*, College of Estate Management, Reading.

RICS (2021). *RICS Rules of Conduct*, RICS.

RICS (2000). *Professional Ethics Guidance Note, RICS Professional Regulation and Consumer Protection Department*, RICS.

RICS (1998). *Agenda for Change*, RICS.

RICS (2017). *RICS Professional Standards and Guidance, Global Conflicts of Interest*, 1st edn. RICS.

RICS (2020). *Futures Report 2020*, RICS.

SCCE (2020). *Compliance and the COVID-19 Pandemic*, Society of Corporate Compliance and Ethics.

Tisser, M. et al. (1996). Chartered surveyors: An international future, *Chartered Surveyor Monthly*, Oct. 15.

Wadlow T. (2020). *YouGov:97% of UK Construction Firms Admit Money Laundering Risk*, Construction.

6   Delivering added value

It is a sad fact that for many clients construction projects fail to achieve a satisfactory conclusion due to a number of factors including the following:

1   Cost over-runs. Historically the construction industry has a poor reputation for delivering projects on budget. This is usually passed off with statements such as every building project is bespoke and carried out in varied and sometimes difficult conditions (adverse weather for example) which makes hitting budget very difficult
2   Unrealistic programmes/schedules. Along with cost overruns, the other curse of construction projects in finishing late. This again is often explained away by the excuses given for cost over runs
3   Not meeting the expectations. Sub-optimal project performance is common in construction projects when the client's perception and the design team's perception of the finished project differ. This can be down to; just plain arrogance on behalf of the client's professional advisors, thinking that they know better than the client, failures to understand the need for the project or problems in the briefing process
4   Linked with the above is failure to clearly define the scope of the project and convey that to the other members of the project team. It is essential at the start of the project to establish the parameters of the project and to convey this to the rest of the project team
5   Lack of definition may also result in the project team being unclear on what must be achieved
6   The construction process is one that is subject to a variety of risks, from adverse ground conditions to shortage of materials. These risks must be managed/mitigated to achieve a successful conclusion to the project, failure to do this could prove disastrous
7   Signature building projects with unfamiliar technology will present a greater challenge to the project manager and team than those that use traditional or well-known construction techniques
8   Lack of a robust business case and commitment from the project sponsors

DOI: 10.1201/9781003293453-6

9   Finally, the position of the project manager is important; not only should the project manager have the responsibility but also the authority to carry through the project.

In an attempt to mitigate the above and deliver added value for construction clients, the quantity surveyor is able to employ the following techniques:

- Project management
- Value Engineering
- Supply chain management
- Risk management.

Project management

Project management is the professional discipline that ensures that the management function of project delivery remains separate from the design/execution functions of a project. It could be thought that the main attributes of a project manager are the so-called hard skills such as financial analysis and technical know-how, although equally important is effective leadership and the ability to communicate and co-ordinate effectively. Project management often necessitates the development of bespoke solutions. The RICS has its own set of project management competencies and APC route.

What is a project?

A project can be thought of a temporary group activity designed to produce a unique product, service, or result, in the case of construction, a new or refurbished construction project, a new piece of infrastructure, etc. Importantly, a project is temporary, in that it has a defined beginning and end and therefore defined scope and resources. Any activities or processes outside of the project scope are deemed to be 'business as usual' and therefore not part of the project.

What is project management?

According the RICS the most important skills required by construction project managers are:

- The supervision of others
- Leadership
- The motivation of others and
- Organisational skills.

Construction project managers are concerned with:

- Developing the project brief
- Selecting, appointing and coordinating the project team

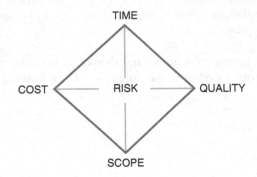

Figure 6.1 Project constraints.

- Represent the client throughout the full development process and
- Managing the inputs from the client, consultants, contractors and other stakeholders.

Typically, project managers will be appointed at the beginning of a project and will assist the client in developing the project brief and then selecting, appointing and coordinating the project team. The project manager will usually represent the client throughout the full development process, managing the inputs from the client, consultants, contractors and other stakeholders. Project management is all about setting and achieving reasonable and attainable goals. It is the process of planning, organizing, and overseeing how and when these goals are met. Unlike business managers who oversee a specific functional business area, project managers orchestrate all aspects of time-limited, discrete projects. During the 1980s, the Ethics, Standards and Accreditation project of the Project Management Institute established the three constraints of project management: time, cost and quality. In addition to project time and cost management a third function – quality was added to be followed eventually by a fourth, scope as illustrated in Figure 6.1. Some project managers add a fifth constraint, risk.

Scope

The scope can be clarified by defining the 'what' of the project as follows. What:

- Will you have at the end of the project?
- Other deliverables could sensibly be carried out at the same time?
- Is specifically excluded from the project?
- Are the gaps or interaction with other projects?
- Is the chance the scope of the project will creep?
- Assumptions must be made?
- Significant difficulties must be overcome?
- Specific conditions or constraints have been stipulated by the client?

Cost

- The cost constraint could be defined in terms of the cost limit or budget for the project.

Time

- The time constraints could be defined as the time to complete the project from access to the site as entered into the contract.

Quality

- The project should result in a functionally efficient building. Quality is all about the extent that something is fit for the purpose for which it is intended. Value engineering can be used to help achieve this.

Risk

- Monitor the progress of the project according to the project plan and the above variables, deal with issues as they arise during the project, look for opportunities to reduce costs & speed up deliver time, and plan, delegate, monitor and control.

The activities that are most commonly involved with construction project management are:

- Identifying and developing the client's brief
- Leading and managing project teams
- Identifying and managing project risks
- Establishing communication and management protocols
- Managing the feasibility and strategy stages
- Establishing the project budget and project programme
- Co-ordinating legal and other regulatory consents
- Advising the selection/appointment of the project team
- Managing the integration and flow of design information
- Managing the preparation of design and construction programmes/ schedules and critical path method networks
- Advising on alternative procurement strategies
- Conducting tender evaluation and contractor selection
- Establishing time, cost, quality and function control benchmarks
- Controlling, monitoring and reporting on project progress and
- Administering consultancy and construction contracts.

Project management phases

Generally, the project management process falls into five stages:
Table 6.1 relates these stages to the RIBA Plan of Work.

Table 6.1 RIBA Plan of Work 2020 compared to classic project management stages

RIBA Plan of Work 2020	Classic Project Management Stages
0 Strategic Definition	1. Initiation
1 Preparation & Brief	1. Initiation
2 Concept Design	2. Planning/organisation
3 Developed Design	2. Planning/organisation
4 Technical Design	2. Planning/organisation
5 Construction	3. Executing and Controlling
	4. Monitoring and controlling
6 Handover & Close Out	5. Closing/evaluation
7 In Use	5. Closing/evaluation

1   *Initiating the project*

The first stage of any project it involves putting the resources in place to complete the project successfully and includes:

- Defining the business model
- Aligning project with business needs
- Defining outcomes/skills and resources
- Setting objectives and
- Deciding to proceed with the project.

Initiation of the project involves setting the quality and quantity parameters as well as trying to avoid the pit falls that plague many projects. This stage may take place as part of the feasibility study and many months prior to project moving forward to the next stage.

2.   *Planning/organisation*

Project management has been described as 80% planning and the success of this stage often determines the success of the project overall. The objective of the planning stage is to evaluate and investigate the best way to achieve the expectations of the client and involves the following tasks:

- Organising workload/planning workload/delegation
- Scoping the project
- Drawing up project schedule with key dates
- Defining project objectives
- Defining major deliverables
- Establishing resources
- Carry out a risk analysis and develop a transparent risk management plan
- Decide to proceed with the project.

The stages in developing a project plan are as follows:

- Brainstorm a list of tasks to be carried out to complete the project, this can be carried out in conjunction with the project stakeholders
- Arrange the tasks in approximate order that they will be carried out and convert into an outline plan; give the tasks a reference number or name
- Estimate, based on previous experiences, the length of time to complete each task and establish task interdependencies.

3. *Executing/Implementation*

This is when the project gets carried out (built) and involves:

- Selecting and appointing the resources to deliver the project with a focus on time/cost/quality and quantity and
- Identifying problems and understanding impact.

4. *Monitoring and controlling*

During this phase, the metrics are established to compare planned with actual progress of the project and involves:

- Tracking the progress of the project and writing progress reports
- Overseeing project status review sessions
- Compiling contingency plans
- Managing third parties
- Managing change and
- Managing budgets.

5. *Closing/Feedback/Review*

- Signing off the project
- Project review
- Lessons learnt.

The construction project manager

The aim of project management is to ensure that projects are completed at a given cost and within a planned time scale. Before beginning to examine how a construction project manager operates it is first necessary to take a wider look at generic project management skills and techniques.

The project management life cycle

Project management has many definitions however it may be regarded as; the professional discipline that ensures that the management function of project

Figure 6.2 Project management lifecycle.

delivery remains separate from the design/execution functions of a project and into these generic skills have to be interwoven the specific skills required for construction projects (Figure 6.2).

Any quantity surveyors reading this book will recognise the three of the constraints/objectives discussed previously (Figure 6.1) that need to be controlled by the project manager to deliver project benefits, as those generally referred to when selecting an appropriate procurement strategy.

Project management skills

Generic or soft project management skills

Project management and the project manager are not unique to the construction industry and there are a number of generic project management skills common to all sectors and industries, for example:

- *Human skills* – the ability to communicate with and motivate people
- *Organisational skills* – management of time, information and costs
- *Technical skills* – industry-specific knowledge and expertise.

Project management, then, is the application of knowledge, skills and techniques to execute projects effectively and efficiently. Very often new projects involve change, either in terms of the organisation or to personnel within the organisation and the project manager should be aware of this. Project management and change management (see Chapter 1) are distinct but interwoven techniques (Figure 6.1).

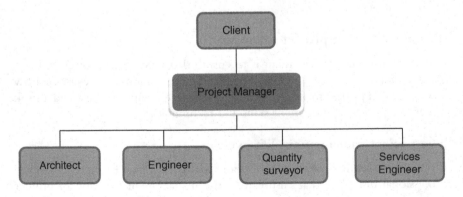

Figure 6.3 Traditional role for project manager in construction project.

Project management tools and techniques

The widespread use of programmes and IT packages during the past 30 years or so has revolutionised the way in which project managers work. Systems such as:

- PRINCE2
- PRIME
- Microsoft Project and
- DefinIT.

are now widely used and the increasing adoption of BIM help the project manager work more efficiently and effectively.

PRINCE2

PRINCE2 or PRojects IN a Controlled Environment is a project methodology developed by the private sector and adapted for use in the public sector originally for use on IT projects. The system is, however, not a software package but can be used on a range of projects from small individual ones to mega projects. Although not in itself a software package there are over 50 tools supporting the methodology. It is not a standard approach and needs to be customised for each project. PRINCE is open access, that's to say free and is used throughout the UK as well as internationally, although it will be necessary to invest in training to get the most out of the system or at least buy the official PRINCE2 official book bundle from the OGC/Cabinet Office.

PRINCE2 identifies six project variables or performance targets:

- *Time;* when will the project finish
- *Cost;* are we within budget
- *Quality;* fit for purpose
- *Scope;* avoid scope creep – uncontrolled change
- *Benefits;* why are we doing this project?
- *Risk;* risk management what happens if?

Another feature of the system are targets for these variables. These targets are set in at the planning stage and regularly checked by the project manager during the project.

Microsoft Project

Microsoft Project (2021) is a project management software programme, developed and sold by Microsoft, which is designed to assist a project manager in developing a plan, assigning resources to tasks, tracking progress, managing the budget and analysing workloads. Microsoft Project was the company's third Microsoft Windows-based application, and within a couple of years of its introduction it became the dominant PC-based project management software.

Newforma

In addition to project management frameworks, there are also a number of other solutions available that can both enable greater efficiency and collaboration between members of the project team. One such solution is Newforma.

Newforma is a system that facilitates project collaboration using the Project Cloud; a web-hosted construction collaboration software that integrates information from the design, construction and owner's team that can be accessed from portable mobile devices. Project information management (PIM) addresses the basic needs of organising, finding, tracking, sharing monitoring and reusing technical project information and communications in a way that is completely aligned with the people and processes that need the information

The role of the project manager and the contract

JCT (16)

The role the project manager is not referred to within JCT (16) nor is there any place within the Articles of the contract to name a project manager, instead the Contract Administrator is referred to as the person with the responsibility of administering, but not necessarily managing the works.

NEC4

NEC uses the term 'project manager' to describe the employer's representative who is tasked with the responsibility of administering the works. In addition to the supervision of the works, the NEC guidance notes proposes that the client/sponsor should appoint a project manager in the early stages of the design sequence to manage the procurement and pre-construction process and not for simply the construction phases and therefore it follows that NEC envisages the project manager's role extending from Stage 1 to Stage 6 in the RIBA Plan of Work 2020.

The role of the project manager under NEC4 can broadly be defined as to:

- Communicate and issue documents in accordance with the contract and generally act in a spirit of trust and co-operation
- Monitor the programme and the contractor's progress against the programme including reviewing and incorporating proposed alterations
- By fairly managing the compensation event process utilising change management

- Assess payments and
- Risk management.

Value analysis/engineering/management

Value engineering/value management (VE/VM) are among the most misused and misunderstood terms in the UK construction industry. It is used by many as a term to describe what has traditionally been termed cost reconciliation – that's to say; making the job fit the price. Nothing could not be further from the concept of value analysis developed by Lawrence Miles in the late 1940s. There are many opportunities to carry out VE/VM studies during a project however the earlier they are introduced then the greater the impact on costs.

Value engineering or value management?

VE and VM are used interchangeably. *The RICS Black Book* guidance notes try to clarify any confusion between the two terms as follows:

- *Whereas the output from a VM study is a report outlining different approaches to the relationship between project objectives and business needs, or to strategic, project-related problems such as which site to select for a new development or which procurement route to use in contrast*
- *The output from a VE study produces a summary of different approaches to achieving the required functionality for a particular material, component or system, the comparative costs of each of the approaches assessed, and a recommended approach that provides the best value for the project.*

Developed first in the United States for the manufacturing and production sectors, by Lawrence D. Miles, in the immediate post–Second World War era as value analysis, later rebadged as VE/VM. For a more objective view of the process, perhaps the reference point should be SAVE, The International Society of American Value Engineers whose definition of VE is:

> *A powerful problem-solving tool that can reduce costs while maintaining or improving performance and quality. It is a function-oriented, systematic team approach to providing value in a product or service.*

Why think in functions?

Buy function, don't buy product

The advantage that VE has over a traditional design review or cost re-conciliation exercise is that VE concentrates on function. The advantages of a function orientated approach are as follows:

- Clarifies intent/purpose
- Helps to achieve consensus among stakeholders
- Broadens individual/group viewpoint(s) by showing the bigger picture
- Aids decision-making as aesthetic issues are not the main focus
- Has been proved to save money
- De-personalises the conversation
- Minimizes emotion.

In essence, VM is concerned with the 'what' rather than the 'how' and would seem to represent the more holistic approach now being demanded by some UK construction industry clients. The basis of VM is to analyse, at the outset, the function of a building, or even part of a building, as defined by the client or end user. Then, by the adoption of a structured and systematic approach, to seek alternatives and remove or substitute items that do not contribute to the efficient delivery of this function, thereby adding value. The golden rule of VE/VM is that as a result of the value process: the function(s) of the object of the study should be maintained and if possible enhanced, but never diminished or compromised.

There is no right and wrong length of time for a workshop or for some of the techniques described below. The traditional approach advocated in North American practice was the use of a five-day workshop (see Figure 6.4) although seldom used in the UK due in part to the resources, both human and physical, needed for a successful outcome. In addition, as these workshops are often conducted by an external VM practitioner there can be problems in getting the outcomes accepted by all the stakeholders. A VE study needs to be facilitated and this can be done either by a VM practitioner or a quantity surveyor with VE training.

The process

> Defining Function
> Classifying Function
> Developing Function Relationships
> Assigning cost to function
> Establishing function worth.

Pre-study

A VE study takes place over several phases see Figure 6.4 and prior to the workshop a briefing document should be prepared and circulated to workshop members containing some or all of the following information:

- Initial and life-cycle cost estimates
- Drawings and/or photographs
- Business Processes

```
┌─────────────────────────────────────┐
│            PRE-WORKSHOP              │
├─────────────────────────────────────┤
│         User/Customer Attitudes.     │
│         Complete Data Files.         │
│         Evaluation Factors.          │
│         Study Scope.                 │
│         Data Models.                 │
│       Determine Team Composition     │
└─────────────────────────────────────┘
```

```
┌───────────────────────────────────────────────┐
│                VALUE WORKSHOP                  │
├───────────────────────────────────────────────┤
│                                                │
│              Information Phase                 │
│           Complete Data Package                │
│               Finalise scope                   │
│                                                │
│           Function Analysis Phase              │
│             Identify Functions.                │
│             Classify Functions.                │
│              Function Models.                  │
│          Establish Function Worth.             │
│               Cost Functions.                  │
│           Establish Value Index.               │
│          Select Functions for Study.           │
│                                                │
│                                                │
│               Creative Phase                   │
│       Create Quantity of Ideas by Function.    │
│                                                │
│              Evaluation Phase                  │
│         Rank & Rate Alternative Ideas.         │
│          Select Ideas for Development.         │
│                                                │
│             Development Phase                  │
│               Benefit Analysis.                │
│           Technical Data Package.              │
│             Implementation Plan.               │
│               Final Proposals.                 │
│                                                │
│                                                │
│             Presentation Phase                 │
│              Oral Presentation.                │
│               Written Report.                  │
│       Obtain Commitment for Implementation     │
│                                                │
└───────────────────────────────────────────────┘
```

```
┌─────────────────────────────────────┐
│            POST WORKSHOP             │
├─────────────────────────────────────┤
│          Complete Changes.           │
│          Implement Changes.          │
│           Monitor Status.            │
└─────────────────────────────────────┘
```

Figure 6.4 Traditional VE phases.

- Staffing
- Stakeholder input
- Design reports
- Existing and proposed solutions
- Constraints and commitments
- Risk Registers
- Any other information that may help the team to understand the project, process, or product.

1   *Information phase*

Understanding the problem. It is especially important that all those involved in either VM or VE properly understand the scenario being studied and the value criteria. In both VM and VE, this needs input from the key stakeholders. The output of this step is a description of what constitutes success for the project and involves (Figure 6.5):

- Identify functions
- Classify functions
- Build function models
- Establish function worth
- Define cost functions
- Establish value index
- Select functions for study.

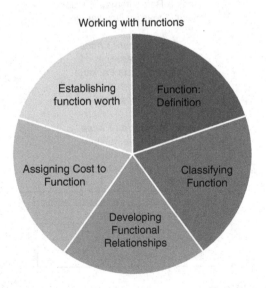

Figure 6.5 Working with functions.

The techniques that can be used to define and analyse functions are as follows:

- Value trees
- Decision analysis matrix
- Functional Analysis System Technique Diagrams (FAST) (see Figure 6.6)
- Criteria scoring.

Once the function of an item (see Figure 6.5) has been defined then the cost or worth can be calculated and the worth/cost ratio scrutinized to determine value for money. Definition of function can be problematic; experience has shown that the search for a definition can result in lengthy descriptions that do not lend themselves to analysis. In addition, the definition of function must be measurable. Therefore, a method has been devised to keep the expression of a function as simple as possible; it is a two-word description made up from a verb and a noun. For example, gather information, triage patients.

Classifying functions

Functions are not of equal importance and are classified as:

- Primary or basic and
- Secondary or support.

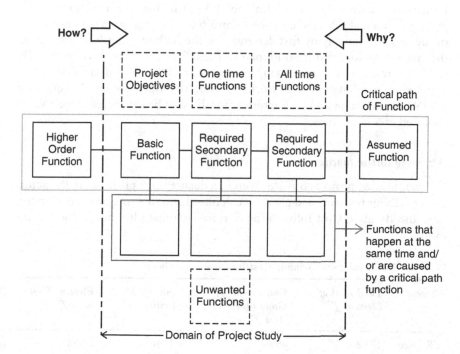

Figure 6.6 FAST diagram.

Basic functions or needs are functions that make the project or service work, if omitted it would impact on the effectiveness of the completed project. Out of the list of basic functions emerges the highest order function, which can be defined as the overall reason for the project and meets the overall needs of the client. If using a FAST diagram (see Figure 6.6), the highest order function is placed to the left of the vertical scope line. The second grouping of functions, the so-called supporting functions, may in majority of cases contribute nothing to the value of a building. Supporting functions generally fall into the following categories:

- Assure dependability
- Assure convenience
- Satisfy user
- Create image.

At first glance these categories may seem to have little relevance to a construction related activities, until it is understood that for example, the 'create image' heading includes items such as aesthetic aspects, overall appearance, decoration and implied performance, such as reliability, safety, etc. Items that in themselves that are not vital for the integrity of the project, but nevertheless may be high on the client's list of priorities.

Developing functional relationships

Functional Analysis System Technique (FAST) models are a method of depicting functional relationships (see Figure 6.6). The model works both vertically and horizontally by first determining the highest order function, called the task or mission, that is positioned to the left of the vertical scope line. By working from the left and asking the question HOW? and employing the verb/noun combination and working from the right asking the question WHY? the functions and their interrelationship can be mapped and their value allocated at a later phase.

Assigning cost to function

Conventionally, project costs are given in a detailed cost plan, where the actual costs of labour materials and plant are calculated and shown against an element, as in the Building Cost Information Service's standard list. See Table 6.2 for example.

Table 6.2 Elemental costs (Building Cost Information service)

Element	Total Cost of Element £	Cost Per m² of Gross Floor Area £	Element Unit Quantity	Element Unit Rate £
5. Services	430,283	45.00	8025 m²	53.62

VE is based on the concept that clients buy functions, not materials or building components, as defined and expressed by their user requirements. Therefore, splitting costs among the identified functions, such as a FAST diagram shows how resources are spent to fulfill these functions. Costs can then be viewed from the perspective of how efficiently they deliver the function. Obviously, the cost of each element can cover several functions, for example, the BCIS element 5. Services may contribute to the delivery of several functions of the project. It is therefore necessary at the outset to study the cost plan and to allocate the costs to the appropriate function (see Table 6.3).

Establishing function worth

The next step is identification of the functions that contain a value mismatch, or in other words seems to have a high contribution to the total project cost in relation to the function that it performs, following on from this, the Creative Phase will concentrate on these functions. Worth is defined as 'the lowest overall cost to perform a function without regard to criteria or codes'. Having established the worth and the cost the value index can be calculated. The formula is Value = Worth/Cost. The benchmark is a ratio of 1.0.

A similar exercise is carried out until all the project costs are allocated to functions.

Table 6.3 Example of cost allocation to function for new GP practice

Element: 5 Services Function	Elemental Cost: £430,283 Cost £
3.1 Identify patients	12,000
3.2 Maintain records	30,000
3.3 Study diagnosis	25,900
3.4 Increase availability	6,900
3.5 Maintain hygiene	7,000
4.1 Refer patient	45,000
4.2 Treat patient	56,000
4.3 Process records	40,000
4.4 Circulate people	60,889
5.1 Council patient	26,605
5.2 Reduce stress	4,989
5.3 Protect privacy	38,000
6.1 Comfort patient	34,000
6.2 Appear professional	43,000
	£430,283

The FAST diagram illustrated in Figure 6.6 is characterised by the following:

- The vertical 'scope line' which separates and identifies the highest-level function – the task or mission – from the basic and supporting functions. It is pivotal to the success of a functional analysis diagram that this definition accurately reflects the mission of the project
- The division of functions into Needs or Basic Functions – with these functions the project will not meet client requirements and Wants or Supporting Functions; these are always usually divided into the four groups previously discussed. The project could still meet the client's functional requirements if these wants are not met or included
- The use of verb/noun combinations to describe functions
- Reading the diagram from the left and asking the question how the function is fulfilled, provides the solution
- Reading the diagram from the right and asking the question Why? identifies the need for a particular function
- The right-hand side of the diagram allows the opportunity to allocate the cost of fulfilling the functions in terms of cost and percentage of total cost.

Therefore, the FAST diagram in Figure 6.6 clearly shows the required identified functions of the project, together with the cost of providing those functions.

2.   *Creative phase*

What now follows is the meat of the workshop – a creative session that relies on good classic brainstorming of ideas, a process that has been compared by those who have experienced it to a group encounter session, the aim of which is to seek alternatives. The discussion may be structured or unstructured – Larry Miles was quoted as saying that the best atmosphere to conduct a study was one laced with cigarette smoke and Bourbon, but in these more politically correct times, these aids to creativity are seldom employed. The rules are simple. Nobody is allowed to say, 'That won't work'. Anybody can come up with a crazy idea. These sessions can generate hundreds of ideas, of which perhaps 50 will be studied further in the workshop's Evaluation Phase. Those ideas will be revisited and some discussion will take place as to their practicality and value to the client. Every project will have a different agenda. Next the process moves onto identifying different solutions to provide the necessary functionality or meet the objectives requires creativity and open thinking. This is the step that construction professionals often find the most challenging. However, there are many well-understood techniques that have been used for creative problem-solving whichever technique is used, the emphasis is on generating as many ideas and potential solutions as possible without critiquing or reviewing them at this point. This step normally requires a workshop approach, so surveyors are likely to be involved as part of the professional team.

3.   *Evaluation phase*

This stage will involve the quantity surveyor as part of the evaluation process and will consider estimates of cost of the various solutions. This step takes each of the technical ideas generated and assesses them against the value criteria discussed during the information gathering stage. A shortlist will be drawn up. Some elimination suggestions will be straightforward on the grounds of impracticability. The remaining solutions are then analysed in greater detail again with the quantity surveyor being called on to advise on the costs of alternatives. Part of this stage should also involve an analysis of life cycle costs and a risk assessment. It is suggested by the RICS Black Book guidance on VE/VM that the selection of the best ideas to be further developed could be based on:

• Open discussion
• Open voting after a discussion and
• Secret voting after a discussion.

The best of the recommendations is then fully developed by the team, typically on day four of the workshop and studies are carried out into costs and life cycle costs of a proposed change before presentation to the client on the final day. It is an unfortunate fact of life of the classic five-day workshop that the team member tasked with costing the recommendations, usually the quantity surveyor, must work into the night on the penultimate day. Finally, a draft report is approved and a final report is written by the team leader. In addition to the above procedures, risk assessment can, or as is thought in some circles should, be introduced into the process. As the value analysts go through and develop value recommendations, they can be asked to identify risks associated with those recommendations, which can either be quantitative or qualitative. And if brainstorming sounds just a little esoteric to the quantity surveying psyche, take heart the results of a VE workshop usually produce tangible results that clearly set out the costs and recommendations in a very precise format.

4.   *Development phase*

Identify the best solution and make recommendations The identified solutions are reduced to a clear favourite, or perhaps a very small number of equally good solutions, through one or more rounds of shortlisting. The outcome of this step and of the study is the description of the preferred solution or solutions. Develop ideas to enough detail that they can be compared against the original solution and provide the opportunity for decision makers to question team and assess recommendations. The process may include the following:

• Sketches of the proposed solution(s)
• Cost breakdown in elemental or elemental groups
• Performance measures

- Risks
- Advantages and disadvantages.

This will usually be in the form of a report that also documents the whole VE process and may include alternative solutions.

5. *Presentation phase*

Present the finding of the study to the client in the form of a report and/or presentation and securing the agreement of the stakeholders.

Some or all the above stages may take place as part of a VE workshop although some stages for example, gathering information and costing alternative solutions, can equally well be done outside of a workshop as a desk top study.

Using VE improved value may be derived by:

- Providing for all required functions, but at a lower cost
- Providing enhanced functions at the same cost
- Providing improved function at a lower cost – the Holy Grail.

The question is often asked; are there projects that are beyond VM? The answer is most certainly – yes. There are many high-profile examples that flaunt the drive to lean construction and these mainly fall into the category of projects for which making a statement either, commercially, politically, or otherwise as their primary their highest order function. Flyvbjerg in his book *Megaprojects and Risk: An Anatomy of Ambition* cites several examples of international megaprojects that have developed their own unstoppable political momentum.

Supply chain management

A construction project team is traditionally a temporary group designed and assembled for the purpose of the project. It is made up by different companies and practices, which have not necessarily worked together before and which are tied to the project by means of varying contractual arrangements. This is what has been termed a temporary multi-organisation; its temporary nature extends to the work force, which may be employed for a particular project, rather than permanently. These traditional design team/supply chain models are the result of managerial policy aimed at sequential execution and letting out the various parts of the work at apparently lowest costs. The problems for process control and improvement that this temporary multi-organisation approach produces are related to:

- Communicating data, knowledge and design solutions across the organization
- Stimulating and accumulating improvement in processes that cross the organisational borders

- Achieving goal congruity across the project organisation and
- Stimulating and accumulating improvement inside an organization with a transient workforce.

Most of what is encompassed by the term supply chain management was formerly referred to by other terms such as 'operations management' but the coining of a new term is more than just new management speak, it reflects the significant changes that have taken place across this sphere of activity. These changes result from changes in the business environment. Most manufacturing companies are only too aware of such changes, increasing globalisation, savage price competition, increased customer demand for enhanced quality and reliability etc. Supply chain management was introduced in order that manufacturing companies could increase their competitiveness in an increasingly global environment as well as their market share and profits by:

- Minimising the costs of production on a continuing basis
- Introducing new technologies
- Improving quality
- Concentrating on what they do best.

The contrast between traditional approaches and supply chain management can be summed up as follows in Figure 6.7.

Unlike other market sectors, because most organisations working in construction as small, the industry has no single organization to champion change. When a series of Government reports call for a 30%–50% reduction in costs, the knee jerk response from some quarters the profession and industry was that cost = prices and therefore it was impossible to reduce the prices entered in the bill of quantities by this amount, therefore the target was unrealistic and unachievable. But, reducing costs goes far beyond cutting the prices entered in the bill of quantities, if it ever did, it extends to the re-organisation of the whole construction supply chain to eliminate

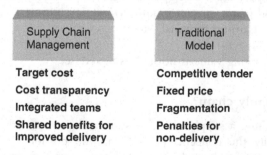

Supply Chain Management	Traditional Model
Target cost	**Competitive tender**
Cost transparency	**Fixed price**
Integrated teams	**Fragmentation**
Shared benefits for Improved delivery	**Penalties for non-delivery**

Figure 6.7 The contrast between traditional approaches and supply chain management.

waste and add value. The immediate implications of supply chain management are:

- Key suppliers are chosen on criteria, rather than job by job on competitive quotes
- Key suppliers are appointed on a long-term basis and pro-actively managed and
- All suppliers are expected to make sufficient profits to re-invest.

How many project managers have asked themselves this question at the outset of a new project: '*What does value mean for my client?*' In other words, in the case of a new plant to manufacture say, pharmaceutical products; what is the form of the built asset that will deliver value for money, over the life cycle of the building for a client? For many years, whenever clients have voiced their concerns about the deficiencies in the finished product, all too often the patronising response from the profession has been to accuse the complainants of a lack of understanding in either design or the construction process or both. The answer to the value question posed earlier will of course vary between clients, a large multi-national manufacturing organisation will have a different view of value to a wealthy individual commissioning a new house, but it helps to illustrate the revolution in thinking and attitudes that must take place. In general, the definition of value for a client is; 'design to meet a functional requirement for a through-life cost'. Project teams are increasingly developing better client focus, because only by knowing the ways in which a particular client perceives or even measures value, whether in a new factory or a new house, can the construction process ever hope to provide a product or service that matches these perceptions. Once these value criteria are acknowledged and understood there are several techniques available in order to deliver to their clients a high degree of the feel-good factor. For example:

- *Measure productivity* – for benchmarking purposes
- *Measure value* – demonstrating added value
- *Measure out-turn performance* – not the starting point
- *Measure supply chain development* – are suppliers improving as expected?
- *Measure ultimate customer satisfaction* – customers at supermarket, passengers at airport terminal, etc.

Of course, measuring value is not an easy exercise.

What is a supply chain?

Before establishing a supply chain or supply chain network, it is crucial to understand fully the concepts behind and the possible components of, a complete and integrated supply chain. The term supply chain has become used to describe the sequence of processes and activities involved in the complete manufacturing and distribution cycle – this could include everything from product design through materials and component ordering through manufacturing and assembly until the finished product is the hands of the final owner. Of course, the nature of the supply chain varies from industry to industry. Members of the supply chain can be referred to as upstream and downstream supply chain members as shown in (Figure 6.8). Supply chain management, which has been practiced widely for many years in the manufacturing sector, therefore refers to how any manufacturer involved in a supply chain, manages its' relationship both up and down stream with suppliers to deliver cheaper, faster and better. In addition, good management means creating a safe commercial environment, in order that suppliers can share pricing and cost data with other supply team members.

The more efficient or lean the supply chain the more value is added to the finished product. As if to emphasis the value aspect some managers substitute the word value for supply to create the value chain. In a construction context supply chain management involves looking beyond the building itself and into the process, components and materials which make up the building. Supply chain management can bring benefits to all involved, when applied to the total process which starts with a detailed definition of the client's business needs which can be provided using VM and ends with the delivery of a building which provides the environment in which those business needs can be carried out with maximum efficiency and minimum maintenance and operating costs. In the traditional methods of procurement, the supply chain does not

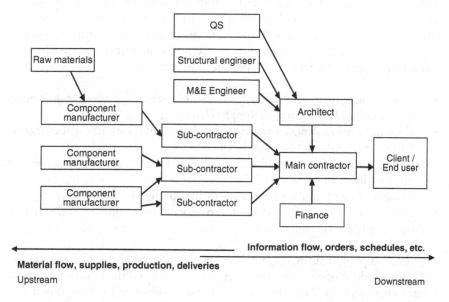

Figure 6.8 Traditional supply chain.

understand the underlying costs, hence suppliers are selected by cost and then squeezed to reduce price and whittle away profit margins.

- Bids based on designs to which suppliers have no input hence no buildability
- Low bids always won
- Unsustainable – costs recovered by other means
- Margins low, so no money to invest in development
- Suppliers distant from final customer so took limited interest in quality.

The traditional construction project supply chain can be described as a series of sequential operations by groups of people or organisations.

Supply chains are unique, but it is possible to classify them generally by their stability or uncertainty on both the supply side and the demand side. On the supply side, low uncertainty refers to stable processes, while high uncertainty refers to processes which are rapidly changing or highly volatile. On the demand side, low uncertainty would relate to functional products in a mature phase of the production life cycle, while high uncertainty relates to innovative products. Once the chain has been catagorised the most appropriate tools for improvement can be selected.

The construction supply chain is the network of organisations involved in the different processes and activities that produce the materials, components and services that come together to design, procure and deliver a building. Traditionally it is characterised by lack of management, little understanding between tiers of other tiers functions or processes and lack of communications and a series of sequential operations by groups of people who have no concern about the other groups or client. Figure 6.8 illustrates part of a typical construction supply chain, although in reality many more sub-contractors could be involved. The problems for process control and improvement that the traditional supply chain approach produces are related to:

- The various organisations come together of a specific project at the end of which disband to form new supply chains
- Communicating data, knowledge and design solutions across the organizations that make up the supply chain
- Stimulating and accumulating improvement in processes that cross the organisational borders
- Achieving goals and objectives across the supply chain and
- Stimulating and accumulating improvement inside an organisation that only exists for the duration of a project.

However, supply chain management takes a different approach that includes:
Prices are developed and agreed, subject to an agreed maximum price with overheads and profit is ring fenced. All parties collaborate to drive-down cost and enhance value with, for example, the use of an incentive scheme. With

cost determined and profit ring fenced, waste can now be attacked to bring down price and add value with an emphasis on continuous improvement.

• As suppliers account for 70%–80% of building costs, they should be selected on their capability to deliver excellent work at competitive cost
• Suppliers should be able to contribute new ideas, products and processes.
• Build alliances outside of project
• Work should be managed so that waste and inefficiency can be continuously identified and driven out.

The philosophy of integrated supply chain management is based upon defining and delivering client value through established supplier links that are constantly reviewing their operation to improve efficiency. There are now growing pressures to introduce these production philosophies into construction and it is quantity surveyors with their traditional skills of cost advise and project management who can be at the fore front of this approach. For example, the philosophy of Lean Thinking, which is based on the concept of the elimination of waste from the production cycle, is of particular interest in the drive to deliver better value. In order to utilise lean thinking philosophy, the first hurdle that must be crossed is the idea that construction is a manufacturing industry which can only operate efficiently by means of a managed and integrated supply chain. At present most clients are required to procure the design of a new building separately from the construction, however as the subsequent delivery often involves a process where sometimes as much as 90% of the total cost of the completed building is delivered by the supply chain members there would appear to be close comparisons with say the production of a motor car or an airplane.

The basics of supply chain management can be said to be:

1   Determine which are the strategic suppliers, and concentrate on these key players as the partners who will maximize added value
2   Work with these key players to improve their contribution to added value and
3   Designate these key suppliers as the 'first tier' on the supply chain and delegate to them the responsibility for the management of their own suppliers, the 'second tier' and beyond.

To give this a construction context, the responsibility for the design and execution of say mechanical installations could be given to a 'first tier' engineering specialist. This specialist would in turn work with its' 'second tier' suppliers as well as with the design team to produce the finished installation. Timing is crucial as first tier partners must be able to proceed confident that all other matters regarding the interface of the mechanical & engineering installation with the rest of the project have been resolved and that this element can proceed independently. Although at least one food retail organisation using supply chain management for the construction of its stores still places the emphasis on the tier partners to keep themselves up to date with progress on the other tiers, as any other approach would be incompatible with rapid timescales that are demanded.

Despite the fact that on the face of it, certain aspects of the construction process appear to be a prime candidate for this approach, the biggest obstacles to be overcome by the construction industry in adopting manufacturing industry style supply chain management are as follows:

1   Unlike manufacturing the planning, design and procurement of a building is at present separated from its' construction or production
2   The insistence that unlike an airplane or motorcar, every building is bespoke, a prototype, and therefore is unsuited to this type of model or for that matter any other generic production sector management technique. This factor manifests itself by:

- Geographical separation of sites that causes breaks in the flow of production
- Discontinuous demand and
- Working in the open air, exposed to the elements. Can there by any other manufacturing process, apart from shipbuilding that does this?

3   Reluctance by the design team to accept early input from suppliers and sub-contractors and unease with the blurring of traditional roles and responsibilities.

There is little doubt that the first and third hurdles are the result of the historical baggage outlined in Chapter 1 and that, given time, they can be overcome, whereas the second hurdle does seem to have some validity despite statements from the proponents of production techniques buildings are not unique and that commonality even between apparently differing building types is a high as 70%. Interestingly though, one of the main elements of supply chain management, Just in Time (JIT) was reported to have started in the Japanese Shipbuilding Industry in the mid 1960s, the very industry that opponents of JIT in construction quote as an example where, like construction, supply chain management techniques are inappropriate. Therefore, the point at which any discussion of the suitability of the application of supply chain management techniques to building must start with the acceptance that construction is a manufacturing process, which can only operate efficiently by means of a managed and integrated supply chain. One fact is undeniable – at present the majority of clients are required to procure the design of a new building separately from the construction. Until comparatively recently international competition, which in manufacturing is a major influencing factor, was relatively sparse in domestic construction of major industrialized countries.

Risk management

Risk management is an essential discipline for informed decision-making within all organisations; however, the construction industry's reputation when it comes to managing risk is not impressive. The traditional approach has been to transfer risk down the supply chain like a hot potato. The problem with this approach is that the risk ends up in the lap of a member of the supply chain who is ill equipped to mitigate or manage it. Certain clients remain wedded to forms of contract, some of which are outdated, that they perceive allocate large amounts of risk to the contractor. More than this some clients demand alterations to the standard forms of contract to further off load risk to the contractor and subcontractors. A survey of CBI members found that in an analysis of 10 recent contracts across the public and private sectors found that each contained on average 87 amended or additional clauses, which typically extended each contract by 37 pages, more than doubling the contract length including a JCT contract with an additional 34 pages of amendments. *CBI, Fine margins, Delivering financial sustainability in UK construction February 2020.*

Some of the features and variables which create risks in the construction process are as follows:

- The bespoke nature of the produce
- Over the years buildings have become more complex in nature and it predicted that this trend is set to continue
- The potential for unforeseen circumstances, for example: - unexpected ground conditions, unpredicted weather conditions, a shortage of resources, a new design that proves impossible or very difficult to construct. Only ship building must operate under similar circumstances
- The length of the contract (projects vary in the time required for completion from days to years). The longer the contract period the greater the opportunity for risk to occur
- The challenge is exacerbated by the hugely varied nature of risk that differs for every project, which could include anything from the following:

    - Geological conditions of construction site
    - Accuracy and specificity of client brief
    - Collapse of supplier businesses
    - Political factors
    - Fluctuating cost and/or availability of materials
    - Labour costs
    - Site accidents
    - Bad weather.

The traditional cavalier attitude to risk is clearly demonstrated in the collapse of the contractor Carillion in 2018. Based on a projection of strong cash flows (£4.6 billion in 2016), the contractor won implausibly low bids for contracts where risks were ignored. The National Audit Office's report into the collapse of Carillion highlighted several multimillion pounds projects where; risks were ignored, there were basic flaws in design and build contracts and bids were very low and contained inadequate resources.

For the quantity surveyor, a major advantage of resilient risk management is a reduction in unexpected or unplanned problems (fighting fires) that can disrupt working patterns, delay contract completion and increase costs. From the client's point of view, uncertainty around the nature and impact of risks on a project may result in the inclusion of large contingency allowances or the need for insurances or bonds to cover unknowns. Any risk which serves to increase the costs and thereby reduce the 'added value' over the life of the construction project will adversely impact upon the successful outcome of the project.

In the public sector, the lack of risk analysis and the assumption that everything will go according to plan lead to the identification of so-called Over-optimism or Optimism bias as identified in a National Audit Office report (Over-optimism in Government projects, 2013, National Audit Office.) The report investigated *'a particularly persistent risk management problem – the difficulties caused for government projects by unrealistic expectations and over-optimism'* (Figure 6.9).

ISO 31000 helps organisations develop a risk management strategy to effectively identify and mitigate risks, thereby enhancing the likelihood of achieving their objectives and increasing the protection of their assets. Its overarching goal is to develop a risk management culture where employees and stakeholders are aware of the importance of monitoring and managing risk (Figure 6.10).

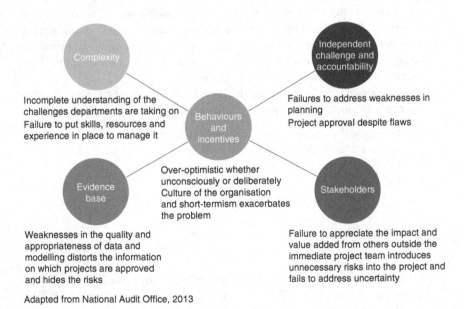

Adapted from National Audit Office, 2013

Figure 6.9 Factor contributing to over optimism.

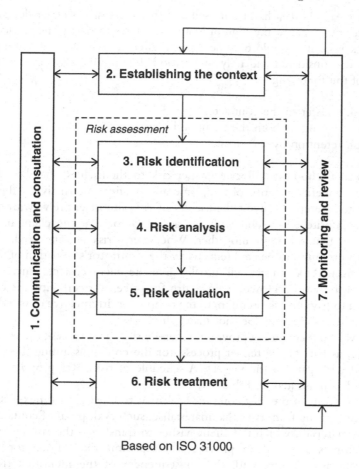

Based on ISO 31000

Figure 6.10 Risk management.

In a report by Mott MacDonald (2002), the main risk areas for public sector projects in UK public sector construction were identified as follows:

- Inadequate business case – 58%
- Disputes and claims – 16%
- Economics (macro-economic business cycle) – 13%
- Complexity of project – 11%
- Poor contractor/ management team – 10%.

Recent examples of a lack of analysis are an arithmetical error in the Department for Communities and Local Government's modelling for the New Homes Bonus led to an over estimate of the impact of the project resulting in the number of new homes to be built being revised from 140,000 to 108,000. When the business case for HS2 was examined in became clear that ten-year old data had been used and that forecast passenger demand did not take into account the high premium rate fares for using the service.

A risk response should only be decided after its possible causes and effects have been considered and fully understood. It will take the form of one or more of the following:

- Risk transfer to the contractor
- Risk sharing by both the client and contractor
- Risk retention by the client.

Clients may choose to allocate greater risk to themselves, thereby deriving potential benefits in terms of cost, time and quality. Alternatively, they may choose to attempt to divest themselves of risk almost entirely, with the resultant opposite effect, and with the added factor that the client may lose control of that risk item altogether. Whenever a risk is transferred, there is usually a premium to be paid (effectively the contractor's valuation of the cost of the risk). Risk transfer will usually give the client cost certainty for that aspect of the works. However, in return for price/cost certainty, the client is required to pay the risk premium to the contractor, irrespective of whether the transferred risk does or does not materialise.

NRM2 now contains a provision whereby the client can ask the contractor to price, as part of the tender process, for the cost of assuming the risks associated with parts of the project. A schedule of risks, if appropriate can be included as illustrated in Table 6.4.

This section of the BQ comprises a list of residual risks (unexpected expenditure arising from risks that materialise, such as disposal of contaminated ground material), which the client wishes to transfer to the contractor. The contractor is required to provide a lump sum fixed price for taking, managing and dealing with the consequences of the identified risk if it materialises. At the time of preparing a BQ, a quantified schedule of works, or other quantity-based document – whether for a complete building project

Table 6.4 NRM2 Appendix F; Schedule of construction risks

Cost Centre	Risk Description	£/p
R001		
R002		
R003		
R004		
R005		
R006		
	Total risk allowance, exclusive of VAT (carried to main summary)	

or discrete works package – there will still be a number of risks to be managed by the client and their project team. This is called the client's residual risk exposure (or residual risks).

Risks that can be designed out or avoided should have been addressed by this stage of the design development process. However, if time does not permit these risks to be designed out or properly dealt with, the risk should be managed using one of these risk response strategies. Risk transfer to the contractor The objective of transferring risk is to pass the responsibility to another party who is able to control it better. If the risk materialises, the consequences are carried by the other party. Risks that the contractor is required to manage, if they materialise, should be fully described so that the extent of services and/or works the client is paying for is clear. Risks to be transferred to the contractor should be listed in the BQ under the heading 'Schedule of construction risks'. The contractor should be assumed to have made due allowance in their risk allowances for programming, planning and pricing preliminaries. Risk allowances inserted by the contractor should be exclusive of overheads and profit.

Risk sharing by client and contractor occurs when a risk is not entirely transferred and some elements of it are retained by the client. It is important that both the client and the contractor know the value of the risk for which they are responsible. The objective should be to improve control and to reduce or limit the cost of the risk to the client, if it materialises. Risks that are to be shared by both client and contractor will normally be dealt with using provisional quantities, with the pricing risk being taken by the contractor and the quantification risk being taken by the client.

Where risks are to be retained by the client, the applicable risk allowances included in the cost plan will be retained and managed by the client or, if empowered by the client, the project team. Before deciding to retain a particular risk, the client might wish to find out what the premium would be if the contractor were to be paid for resolving the risk if it materialises. The client can then decide whether to pay a premium for a defined scope of work. If the client is content to pay a premium for transferring the risk, it is dealt with as a risk transfer. Risks retained by the client are not necessarily controllable.

It is often considered that risks should be held by the party best able to deal with that risk, and this is a factor that should be taken into account when considering the appropriate procurement strategy.

Risk response

A risk response should only be decided after its possible causes and effects have been considered and fully understood. It will take the form of one or more of the following management actions:

* Reduction (avoidance)
* Transfer

- Retention (including sharing) or
- Insure against the risk.

Tools and techniques

There are several techniques available to identify and manage risk. Three of the more popular are as follows:

- Risk registers
- Scenario analysis
- Scensitivity analysis.

Risk registers

The purpose of risk assessment is to understand and quantify the likelihood of occurrence and the potential impacts on the project outturn. Various analytical techniques are available, but the key features are:

Qualitative assessment – to describe and understand each risk and gain an early indication of the more significant risks. In a qualitative analysis, descriptive terms are used such as 'low impact' and 'high probability'. A descriptive written statement of relevant information about a potential risk should be prepared. Issues to be considered should include the following:

- The stages of the project when it could occur
- The elements of the project that could be affected
- The factors that could cause it to occur
- Any relationship or inter-dependency on other risks
- The likelihood of it occurring
- How it could affect the project.

Quantitative assessment – to quantify the probability of each risk occurring and its potential impact in terms of cost, time and performance. In a quantitative analysis, risks will have values attributed to them, such as 'an impact of £10m' or 'a probability of 65%'.

The likelihood of a risk occurring is given a numerical probability. This is measured on the following scale:

0   = impossible for risk to occur
0.5 = even chance of risk occurring and
1   = risk will occur.

Finally, the analysis of probability and impact are combined into a single risk score. This risk score can be presented quantitatively or qualitatively, with quantitative methods often using colours (green, yellow and red) as well as words (Figure 6.11).

Severity Likelihood	Minimal 5	Minor 4	Major 3	Hazardous 2	Catastrophic 1
Frequent A	[Green]	[Yellow]	[Red]	[Red]	[Red]
Probable B	[Green]	[Yellow]	[Red]	[Red]	[Red]
Remote C	[Green]	[Yellow]	[Yellow]	[Red]	[Red]
Extremely Remote D	[Green]	[Green]	[Yellow]	[Yellow]	[Red]
Extremely Improbable E	[Green]	[Green]	[Green]	[Yellow]	[Red] * [Yellow]

High Risk [Red]
Medium Risk [Yellow]
Low Risk [Green]

* High Risk with Single
Point and/or Common
Cause Failures

Figure 6.11 Colour coding risks.

Scenario analysis

Involves producing three scenarios. A criticism of this technique is that it is somewhat subjective and involves determining the cost of the worst, most likely and best case scenarios along with the probability of occurrence, for example:

- Worst case (Pessimistic) – 15%
- Most likely – 50%
- Best case (Optimistic) – 35%.

Variable-by-variable analysis

This approach analyses uncertainty and isolates the effect of change on one variable at a time. The approach is as follows:

- List all the important factors that can affect the successful outcome of the project
- For each factor define a range of possible values
- Generally, three ranges are proposed; Optimistic, Most likely and Pessimistic
- Calculate the cost-benefit ratios or net present values for each of the ranges.

A variable-by-variable analysis assumes that factors affecting a project do not operate independently of one another. By using software such as Microsoft Excel, it is possible to model various scenarios to answer the 'what if' question. For example, what will be the effect if interest rates were to change from the 10% assumed, during the construction period?

	Optimistic	Most Likely	Pessimistic
	9%	10%	11%
Finance costs on £6,847,500	£308.138	£342,375	£376,613

This process should be repeated for all key drivers

Sensitivity analysis (what if?)

The first stage in the process is to create a given set of scenarios based on critical aspects of the project, for example, cost of finance, cost of materials, etc., and determine how changes to the chosen factors will impact the target variable. The results provides estimate based information on which decisions can then be made. In a sensitivity analysis the calculation can be repeated using different figures to find out how sensitive the results are to change and can determines the impact on profit levels.

Risk register

The first step in developing a risk register is to identify all possible project risks. Risk workshops are held to identify all the risks associated with a project that could have an impact on cost, time or performance of the project. The risks that have been identified then need to be analysed for their probability and impact on the project (such as cost, time and resource).

By way of analysis, it is usual to place the list of risks in various categories that could include some or all of the following:

- Strategic risks, such as failure to obtain planning permission, or client funding problems
- External risks, such as changes in the environment
- Project risks, such as overspends or delays to the programme
- Discovery risks, such as poor ground conditions or the like.

ID	Project	Risk description	Likelihood of the risk occurring	Impact if the risk occurs	Severity *Rating based on impact & likelihood.*	Owner *Person who will manage the risk.*	Mitigating action *Actions to mitigate the risk e.g. reduce the likelihood*	Contingent action *Action to be taken if the risk happens.*	Progress on actions	Status
1		Project purpose and need is not well-defined.	Low	High	medium	DC	Complete a business case if not already provided and ensure purpose is well defined.	Escalate to the Board with an assessment of the risk of runaway costs/never-ending project.	Business case re-written with clear deliverables and submitted to the board for approval.	Open
2		Project design and deliverable definition is incomplete.	Medium	High	medium	Project Sponsor	Define the scope in detail via design workshops with input from subject matter experts.	Document assumptions made and associated risks. Request high risk items that are ill-defined are	Design workshops scheduled.	Open

Figure 6.12 Risk register example.

The identified risks are normally placed into a risk register.

The overall list of risks can be as long or as short as is necessary (and this may well depend upon the complexity of the project). These risks will either be generic (common to most if not all construction projects) or will be specific to the particular construction project in question.

In the analysis of the risks in the risk register, a risk scoring exercise is normally undertaken witheach risk event is scored by reference to a pre-agreed scale of the likelihood of the event occurring. An element of judgment is inevitably needed as the project team may not necessarily be aware of the likelihood of client risk events occurring (Figure 6.12).

Expenditure of risk allowances

The risk register should be managed with costs being moved out of risk allowance into base cost estimate/other cost centre as risks materialise although risk allowances should only be expended to which they relate occur. If previously unidentified risks materialise they should be treated as variations.

Bibliography

Association for Project Management (2013). *Value Management Specific Interest Group White Paper for Discussion*.

BIS Research Paper No. 145 (2013). *Supply Chain Analysis into the Construction Industry: A Report for the Construction Industrial Strategy*.

Cartlidge, D. (2013). *RICS Information Paper Construction Sectors and Roles for the Surveyor*, QS & Construction Professional Group of the Royal Institution of Chartered Surveyors.

CBI (2020). *Fine Margins, Delivering Financial Sustainability in UK Construction 2020*.

Chartered Institute of Building (2014). *Code of Practice for Project Management for Construction Development*, CIOB.

CIC (2007). *Construction Industry Council (CIC) Scope of Services Handbook, Major Works*, Construction Industry Council.

Construction 2025 (2013). *Industrial Strategy: Government and Industry in Partnership*, HM Govt.

EC Harris (2013). *Supply Chain Analysis into the UK Construction Sector*.

Flyvberg, B. (2003). *Megaprojects and Risk: An Anatomy of Ambition*, Cambridge University Press.

MacDonald, M. (2002). *Review of Large Public Procurement in the UK*.

Mosey, D. (2009). *Early Contractor Involvement in Building Procurement: Contracts, Partnering and Project Management*, Wiley-Blackwell.

NAO (2013). *Over-Optimism in Government Projects*, National Audit Office.

RIBA (2022). *RIBA Contracts and Law Report 2022*, RIBA Enterprises.

RICS Research Report (2005). *Beyond Partnering: Towards a New Approach in Project Management*, RICS.

RICS (2009). *Development Management Guidance Note*, RICS.

RICS (2017). *Professional Standards and Guidance, UK Value Management and Value Engineering 1st Edition*, QS & Construction Professional Group of the Royal Institution of Chartered Surveyors.

RICS (2021). *NRM 2 New Rules of Measurement Detailed Measurement for Building Works, 2nd Edition*, Royal Institution of Chartered Surveyors.

Index

Printed in the United States
by Baker & Taylor Publisher Services

Printed in the United States
by Baker & Taylor Publisher Services